Shiwen Tong, Dianwei Qian, Chunlei Huo
Hydrogen-Air PEM Fuel Cell

Also of interest

Electrochemical Storage Materials
From Crystallography to Manufacturing Technology
Meyer, Leisegang, 2018
ISBN 978-3-11-049137-1, e-ISBN 978-3-11-049398-6

Power-to-Gas
Renewable Hydrogen Economy for the Energy Transition
Boudellal, 2018
ISBN 978-3-11-055881-4, e-ISBN 978-3-11-055981-1

Electrochemical Energy Systems
Foundations, Energy Storage and Conversion
Braun, 2018
ISBN 978-3-11-056182-1, e-ISBN 978-3-11-056183-8

Hydrogen Storage
Based on Hydrogenation and Dehydrogenation Reactions of Small Molecules
Zell, Langer, 2018
ISBN 978-3-11-053460-3, e-ISBN 978-3-11-053642-3

Chemical Synergies
From the Lab to In Silico Modelling
Bandeira, Tylkowski (Eds.), 2018
ISBN 978-3-11-048135-8, e-ISBN 978-3-11-048206-5

Shiwen Tong, Dianwei Qian, Chunlei Huo

Hydrogen-Air PEM Fuel Cell

Integration, Modeling, and Control

DE GRUYTER

Authors
Dr. Shiwen Tong
College of Robotics
Beijing Union University
No.97 East Road of Beisihuan
Chaoyang District, 100101 Beijing, China
shiwen.tong@buu.edu.cn

Dr. Chunlei Huo
National Laboratory of Pattern Recognition
Institute of Automation, Chinese Academy of Sciences,
No.95 East Road of Zhongguancun
Haidian District, 100190 Beijing, China
clhuo@nlpr.ia.ac.cn

Dr. Dianwei Qian
School of Control and Computer Engineering
North China Electric Power University
No.2 Beinong Road
Changping District, 102206 Beijing, China
dianwei.qian@ncepu.edu.cn

ISBN 978-3-11-060113-8
e-ISBN (E-BOOK) 978-3-11-060215-9
e-ISBN (EPUB) 978-3-11-060036-0

Library of Congress Cataloging-in-Publication Data
Names: Tong, Shiwen, author. | Qian, Dianwei, author. | Huo, Chunlei, author.
Title: Hydrogen-air PEM fuel cell integration, modeling and control / Shiwen Tong, Dianwei Qian, Chunlei Huo.
Description: Berlin ; Boston : Walter de Gruyter GmbH, [2018] | Includes bibliographical references and index.
Identifiers: LCCN 2018030844 (print) | LCCN 2018040189 (ebook) | ISBN 9783110602159 (electronic Portable Document Format (pdf) | ISBN 9783110601138 (print : alk. paper) | ISBN 9783110602159 (ebook pdf) | ISBN 9783110600360 (ebook epub)
Subjects: LCSH: Proton exchange membrane fuel cells.
Classification: LCC TK2933.P76 (ebook) | LCC TK2933.P76 T66 2018 (print) | DDC 621.31/2429--dc23
LC record available at https://lccn.loc.gov/2018030844

Bibliographic information published by the Deutsche Nationalbibliothek
The Deutsche Nationalbibliothek lists this publication in the Deutsche Nationalbibliografie; detailed bibliographic data are available on the Internet at http://dnb.dnb.de.

Typesetting: Integra Software Services Pvt. Ltd.
Printing and binding: CPI books GmbH, Leck
Cover image: Science Photo Library / Giphotostock

www.degruyter.com

Acknowledgments

I would like to thank Professor Guo-ping Liu from the University of South Wales, United Kingdom. He brought me into the academic circle. It is impossible to have this book without his guidance. I would also like to thank Professor Jianqiang Yi from the Institute of Automation, Chinese Academy of Sciences, for his useful discussion of the relevant contents of the book many years ago. Special thanks to Professor Robert Babuška of Delft University of Technology, Netherlands. His Fuzzy Modeling Toolbox and monographs have benefited me a lot. I also want to thank my wife and my daughter in particular. In the creation of this book, my wife shared most of the housework without regrets. And my beautiful daughter brought me joy and infinite motivation, so that I can smoothly complete the book. Finally, many thanks to all the editors and reviewers who contributed to the publication of this book.

https://doi.org/10.1515/9783110602159-201

Acknowledgements

Preface

Proton exchange membrane (PEM) fuel cells are very promising clean energies that can be used in different areas due to their capabilities of fuel diversification, minimum environmental pollution, low noise, high reliability, and good maintenance. PEM fuel cells exist in an enormous variety of forms. According to the difference of oxides, PEM fuel cells can roughly be categorized into two types, that is, hydrogen–oxygen PEM fuel cell and hydrogen–air PEM fuel cell. Between these two types of fuel cells, their mechanisms and dynamics can be described by a series of complicated equations involving electrochemical reactions, ideal gas equations, Dalton's law of partial pressure, law of conservation of mass, and so on. It can be seen from these equations that hydrogen–air PEM fuel cells are inherently nonlinear systems.

Taking applications and economy into consideration, hydrogen–air PEM fuel cells have different hardware structures from the hydrogen–oxygen PEM fuel cells, which rise up on account of control problems, such as pressure balance between the cathode and the anode, voltage tracking on fluctuations of loads, and underactuated operation in flow system.

The control problem has become one of the central research topics. Most of the previous work in this field considers the process or the control approach separately. This book tries to combine these two parts together. The control methods described in this book are closely related to the specific process, which involves not only simulation but also experiments. To circumvent complex mechanism analysis processes, the SIRMs-based (single-input rule modules) fuzzy control method and data-based control approaches (sliding mode output tracking control method) are also proposed.

As a feature of this book, a lot of MATLAB/Simulink, which is a widely used software in universities or institutes, programs are developed. For better understanding of the source codes, each program block is accompanied by a very simple example. The C-S functions have been provided for real-time implementation.

The book has six chapters. Simulative and experimental programs are mainly included in Chapters 4 and 5.

Chapter 1 provides some preliminary knowledge, including the classification, characteristics, working principle, and control problem of the PEM fuel cell.

Chapter 2 investigates the hardware configuration of the 1 kW hydrogen–air PEM fuel cell. It mainly discusses the overall architecture and the structure of each specific subsystem. Particularly, the NetCon system which is used as a control system and can realize a seamless connection with MATLAB/Simulink is presented.

Chapter 3 focuses on the modeling of the hydrogen–air PEM fuel cell. The models are mainly built by mechanism analysis methods implementing electrochemical, mass conservation, and other chemical or physical principles. Some parameters of the

https://doi.org/10.1515/9783110602159-202

models are obtained by identification. For the convenience of simulation and control, the models are translated into Simulink version. The DC/DC buck converter model is also established in the end of this chapter.

Chapter 4 investigates the control methodology of PEM fuel cell without considering the feedback of load current. This chapter starts with some basic concepts about singular pencil model (SPM). Then, simultaneous estimation problem of parameters and states is discussed by means of SPM. Combined with state feedback predictive control, the cathode pressure is controlled in different configurations of controllers, considering the noise information or not. Due to the complicated nonlinear time-varying essence of the fuel cell process, as for nonlinear control method, this book has also proposed a real-time simplified variable domain fuzzy control to further improve the control performance, which can realize both the real-time control and the accurate control at the same time. Simulative and experimental results are illustrated to support the proposed control methods. Finally, all the source codes are drawn.

Chapter 5 focuses on the control methodology of PEM fuel cell considering the feedback of load current. For the control of flow subsystem, such a system actually formed an underactuated control problem. This book has implemented the fuzzy control based on SIRMs dynamically connected fuzzy inference model and realizes the control of two controlled variables, cathode pressure and air flow, with one manipulated variable, set voltage of air mass flowmeter. In fact, the core idea of solving the control of underactuated system is to turn control problems into optimization issues. This book has developed the C-S function-based online random search optimization method to optimize the dynamic importance degree of the fuzzy logic model. To circumvent the soft characteristic of fuel cell stack output, the voltage tracking control method has also been explored, which involved fuzzy cluster modeling, neural network approximation, and sliding mode control technologies. Simulative and experimental results are illustrated to support the control methods. Finally, all the source codes are given.

Chapter 6 draws some concluding remarks and summarizes open problems for future research.

The book can be used in teaching a graduate-level special-topics course in control applications.

In this book, all the control algorithms and their programs are described separately and classified by the chapter name, which can be run successfully in MATLAB 9.1.0.441655 version or other more advanced versions. If you have questions about algorithms and simulation programs, please feel free to contact Shiwen Tong by E-mail: shiwen.tong@buu.edu.cn.

Beijing
Shiwen Tong

Contents

1 Introduction

With the continuous development of mankind, people have become more dependent on energy. However, traditional fossil energy has been constantly decreasing. Furthermore, fossil fuel releases a lot of carbon monoxide, carbon dioxide, sulfur dioxide, and other harmful gases in the combustion process, which poses a threat to the human living environment. For the sustainable development of mankind, almost all countries in the world are committed to finding and developing new energy sources instead of relying on traditional energy sources. These energy sources include solar, wind, bio-, hydrogen, tidal, geothermal, and nuclear energy. After years of exploration and efforts, people have finally found a new energy structure, that is, generating hydrogen and oxygen by electrolysis of water with the solar, and then producing electric energy by the chemical reactions of hydrogen and oxygen. In the future energy systems, the solar energy will serve as a major primary energy alternative to the current coal, oil, and natural gas, the three major energy sources. Hydrogen is the clean energy for the twenty-first century, as an alternative energy source to the gasoline, diesel, city gas, and other. One of the forms of utilization for the hydrogen energy is the conversion of it into electrical energy through various types of fuel cell devices.

1.1 Overview of fuel cell

In 1839, Grove made the first hydrogen fuel cell in the United Kingdom. He developed a single cell with platinum plating electrode, using hydrogen as fuel, oxygen as oxidant, and diluted sulfuric acid as a liquid electrolyte, and produced electricity successfully. The power generation principle for the hydrogen-oxygen fuel cell has been achieved.

The 1950s was a turning point in the development of fuel cells. Bacon in the University of Cambridge, United Kingdom, conducted a long-term and fruitful research for the hydrogen-oxygen alkaline fuel cells. The main contributions are as follows: first, he proposed a new nickel electrode, using double-hole structure to improve the gas transmission characteristics; second, he proposed new preparation process, using lithium-ion embedded nickel plate pre-oxidation roasting to solve the problem of electrode oxidation corrosion; third, he proposed a new drainage program to ensure the quality of electrolyte work. Thus, a 5 kW alkaline fuel cell was successfully developed, whose life can be 1,000 h. This is the first practical fuel cell. Bacon's achievements have laid the technical idea of modern fuel cells, and encouraged people to try to achieve practical commercialization of fuel cells [1].

The 1960s is an important stage of development for the application of fuel cells in the aerospace industry. The United States launched the manned spacecraft plan in

https://doi.org/10.1515/9783110602159-001

1962. It used the polymer fuel cell made by GE. In 1968, alkaline fuel cell, which is utilized the Bacon's technology, was selected by NASA in the Apollo moon landing program [2].

Since the 1990s, the fuel cell has been undergoing rapid development. The research hotspot is the proton exchange membrane fuel cell [3]. Canadian Ballard company, an international fuel cell industry giant, built the world's first fuel cell plant, and formally put it into production in February 2001 [4].

A fuel cell is a power generation device that directly converts chemical energy stored in fuel and oxidant into electrical energy by an electrode reaction. The greatest advantage of it is that the reaction process does not involve combustion, thus, not subject to the restriction of the Carnot cycle. The energy conversion rate can be up to 60% ~ 80%. The practical efficiency is two to three times more than the ordinary gas turbine. In addition, it has the advantages of fuel diversification, minimum environmental pollution, low noise, good reliability, and good maintenance, and is recognized as the preferred clean and efficient power generation technology in the twenty-first century [5].

1.2 Classification of fuel cell

Fuel cells can be classified by the operating temperature, the source of the fuel, and the use of different electrolytes. [6].

With respect to different working temperatures, fuel cells can be classified into low-temperature fuel cell, medium-temperature fuel cell, and high-temperature fuel cell.

With respect to different fuel sources, fuel cells can be classified into direct fuel cell, indirect fuel cell, and renewable fuel cell.

With respect to the use of different electrolytes, fuel cells can be classified into five categories: alkaline fuel cell, phosphoric acid fuel cell, molten carbonate fuel cell, solid oxide fuel cell, and proton exchange membrane fuel cell. The basic characteristics of these five fuel cells are shown in Table 1.1, and the electrochemical reactions at the anode and cathode are shown in Table 1.2.

1.3 Characteristics of fuel cell

Compared with other energy sources, fuel cells have certain advantages, which are discussed in this section.

First, the energy conversion efficiency is very high in a fuel cell. In theory, the fuel cell can convert 90% of the fuel energy into electricity and heat. Phosphorus fuel cell power generation efficiency is currently close to 37% ~ 42%. Molten carbonate fuel cell power generation efficiency can be more than 60%. Solid oxide fuel cell efficiency is still higher. Moreover, the efficiency of the fuel cell is independent of its

Table 1.1: Basic characteristics of different types of fuel cells.

	Alkaline fuel cell	Phosphoric fuel cell	Molten carbonate fuel cell	Solid oxide fuel cell	Proton exchange membrane fuel cell
Abbreviations	AFC	PAFC	MCFC	SOFC	PEMFC
Anode	Pt/Ni	Pt/C	Ni/Al	Ni/ZrO_2	Pt/C
Cathode	Pt/Ag	Pt/C	Li/NiO	$Sr/LaMnO_3$	Pt/C
Electrolyte	KOH(l)	H_3PO_4(l)	K_2/Li_2CO_3	ZrO_2(s)	Nafino(s)
Conductive ions	OH^-	H^+	CO_3^{2-}	O^{2-}	H^+
Fuel	Pure hydrogen	Hydrogen, natural gas, reforming gas	Natural gas, reforming gas, purify gas	Natural gas, purify gas	Hydrogen, methanol, natural gas, reforming gas
Oxidizing agent	Pure oxygen	Oxygen, air	Oxygen, air	Oxygen, air	Oxygen, air
Corrosive	Strong	Strong	Strong	Weak	No
Operating temp/°C	Room temp~200	100~200	650~700	900~1000	Room temp~200
Specific power/W•Kg^{-1}	35~105	120~180	30~40	15~20	300~3000
Life/h	10,000	15,000	13,000	7,000	100,000
Start time	Several min	Several min	>10 min	>10 min	<5 sec
Power level/KW	1~200	10~10000	Thousands to hundreds of thousands	Hundreds of thousands	1~300
Applications	Aerospace, special ground	Regional power station, field power supply	Regional power station	Regional power station, combined cycle power generation	Electric cars, portable power supplies, small power plants

Table 1.2: Basic electrochemical reactions of various fuel cells.

Fuel cell type	Anode reaction	Ion through the electrolyte	Cathode reaction
AFC	$H_2+2OH^-\rightarrow 2H_2O+2e^-$	H^+	$1/2O_2+2H_2O+2e^-\rightarrow 2OH^-$
PAFC	$H_2\rightarrow 2H^++2e^-$	H^+	$1/2O_2+2H^++2e^-\rightarrow H_2O$
MCFC	$H_2+CO_3^{2-}\rightarrow H_2O+CO_2+2e^-$	CO_3^{2-}	$1/2O_2+CO_2+2e^-\rightarrow CO_3^{2-}$
SOFC	$H_2+O^{2-}\rightarrow H_2O+2e^-$	O^{2-}	$1/2O_2+2e^-\rightarrow O^{2-}$
PEMFC	$H_2\rightarrow 2H^++2e^-$	H^+	$1/2O_2+2H^++2e^-\rightarrow H_2O$

size. Thus, the fuel cell can run at its half rated power while maintaining high fuel efficiency [7].

Second, fuel cells have a high degree of reliability. The fuel cell power generation device is made up of a battery pack connected by a series of cells. The battery pack is connected in parallel to determine the size of the entire power generation unit. These battery combinations are of module structure, constituting the basic operation and maintenance unit of the power generation system. Thus, maintenance becomes very convenient. In addition, even if it is overloaded above the rated power, or below the rated power, it can withstand without lowering the efficiency dramatically [8].

Third, fuel cells have good environmental benefits. The by-product of electrochemical reaction in a fuel cell is water. Since the rotating parts of the fuel cell are rare, the working noise is very low. The environmental friendliness of the fuel cell is the main reason, which makes it a strong vitality and long-term development potential.
But fuel cell still has many shortcomings, which can be summarized as follows:

first, it is costly and expensive;
second, the life and stability are not ideal in a high-temperature environment;
third, the fuel cell technology is not popular enough; and
fourth, there is no perfect fuel cell supply system.

1.4 PEM fuel cell working principle

The proton exchange membrane (PEM) fuel cell is a new type of fuel cell in the developing process. The electrolyte is a solid organic film, which can conduct protons in case of humidification. Platinum is generally used as a catalyst and the working temperature is generally 60°C ~ 80°C.

Each cell in the PEM fuel cell is mainly composed of a membrane electrode, a sealing ring, and a flow field plate with a gas passage. Membrane electrode is the core of the PEM fuel cell. In the middle of the membrane electrode is a layer of thin film – PEM – which does not conduct electrons but is an excellent conductor of hydrogen ions, both as an electrolyte providing hydrogen ion channels and as a diaphragm isolating bipolar reaction gas. At both sides of the membrane are gas electrodes, made of carbon paper and catalyst. The anode is the hydrogen electrode, and the cathode is the oxygen electrode. The flow field plate is usually made of graphite. PEM fuel cell uses hydrogen as fuel and air or pure oxygen as oxidant. Multiple cells are connected in series or in parallel to form a different battery pack (stack).

The working principle of the PEM fuel cell is shown in Figure 1.1.

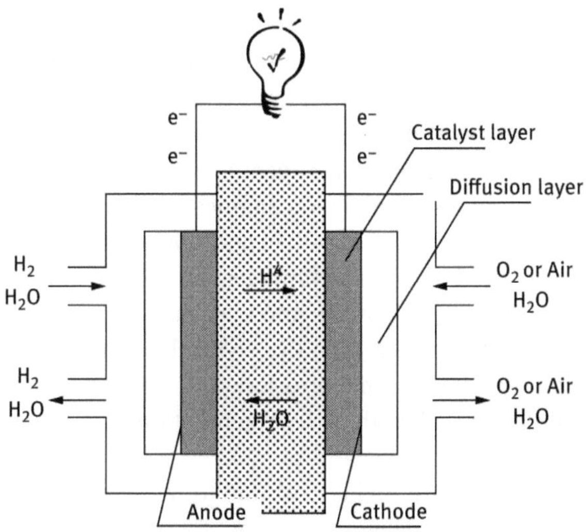

Figure 1.1: PEM fuel cell working principle.

Hydrogen reaches the anode of the battery through the gas passage in the bipolar plate. On the other end of the battery, oxygen or air goes to the cathode of the battery through the gas passage in the bipolar plate. Hydrogen undergoes a chemical reaction under the action of the anode catalyst, and then dissociates into two hydrogen ions, that is, protons, and releases two electrons at the same time. They pass through different paths to the cathode. Electrons reach the cathode through the external circuit rather than the electrolyte, forming current in the process of electronic movement, while the protons reach the cathode through the electrolyte. Oxygen, hydrogen ions, and electrons take a reaction to generate water under the action of the cathode catalyst. The chemical reaction can be expressed by the following chemical equation:

$$\text{Cathode}: \frac{1}{2}O_2 + 2H^+ + 2e^- \rightarrow H_2O$$

$$\text{Anode}: H_2 \rightarrow 2H^+ + 2e^-$$

$$\text{Overall}: \frac{1}{2}O_2 + H_2 \rightarrow H_2O$$

1.5 Characteristics of the PEM fuel cell

The PEM fuel cell has many characteristics such as no corrosion, long life, high specific power, low operating temperature, fast start, wide fuel source, and environment-friendly. It is considered as one of the most promising energy sources [9]. In recent years, PEM fuel cells have made great progress in the applications of electric vehicles, small power plants, and portable power sources. Fuel cell buses and small power plants manufactured by Canadian Ballard company have been cast into actual use. While the German Mercedes-Benz car company, Japan's Toyota electric car company, the Swedish electric car company, and the US General Motors company have produced fuel cell vehicles, some other well-known car companies are also actively committed to the development of fuel cell vehicles. PEM fuel cell is being considered as the fourth generation of power technology and the most promising alternative to automotive internal combustion engines [1].

Compared with other types of fuel cells, PEM fuel cell, in addition to the general characteristics of fuel cells, also has the following advantages: room temperature operation, quick start and close, high specific power; zero emissions, pure water by-product, no pollutant emissions, environment-friendly; no mechanical moving parts, low operating noise, high reliability; no corrosion and easy to make a reversible renewable fuel cell integrated with power generation and water electrolysis.

The shortcomings of the PEM fuel cell are higher cost due to the usage of the platinum group metal as the electro catalyst, higher requirement for the purification of fuel because of the sensitivity for the catalytic activity to the carbon monoxide, lower recoverable waste heat than other types of fuel cells because of the low operating temperature, and complex water and heat management due to the significant effect on the battery performance by the PEM moisture and temperature.

1.6 Research progress for the control method of the PEM fuel cell

In 2002, the researchers from the University of Michigan proposed a nonlinear fuel cell dynamic system model oriented to the control research, taking into account the dynamic characteristics of the flow, the dynamic characteristics of the compressor, and the partial pressure of the reactants. They also designed observer-based feedback and feedforward controller to coordinate the contradiction between the rapid response of the output power and the parasitic power loss of the supercapacitor during rapid load changes [10–12]. In 2004, Pukrushpan designed an air flow controller, based on the dynamic model, to prevent the oxygen starvation in the case of sudden changes in stack current. He also designed a linear optimal controller to find the optimal operating frequency for coordination between stack output power control and stack oxygen starvation control [11],[13].

Variable pressure strategies for fuel cell systems can increase the efficiency of the system. The air compressor is the main energy consumer, especially at low power output. However, variable pressure strategies require rapid and accurate control of mass flow and pressure. When two separate Proportional-Integral (PI) control loops are used for mass flow and pressure control, these two variables interact and cannot achieve a satisfactory control performance. Rodatz et al. implemented a linear quadratic Gaussian regulator not only to achieve the decoupling of mass flow and pressure but also to reduce the dynamic response time [14].

When the fuel cell generates current, the key is that the oxygen consumed in the reaction should be quickly replenished by the air supply system to avoid hypoxia and damage. For stand-alone fuel cells, lack of control over high current requirements so as to protect the battery from hypoxia has become a crucial issue.

Vahidi et al. added an auxiliary power source – supercapacitor – in the fuel cell to avoid hypoxia. To keep the system from getting damaged due to operational constraints, Vahidi et al. put forward a diversion strategy, which was to divide the power between the battery and the auxiliary source through the constrained optimization structure such as model predictive control. The simulation results showed that the reactants deficit could be reduced from 50% in the original independent battery structure to 1% in the current hybrid system when the power in the stack suddenly increased [15],[16].

Caux et al. proposed the discrete form of the state space models for the fuel cell flow and pressure, and designed state feedback controllers to achieve flow and pressure control by using pole placement method [17]. Chen et al. designed a variable digital input two-loop Proportional-Integral-Derivative (PID) controller which achieved the balance control of the anode and cathode pressur of the PEM fuel cell [18].

Almeida et al. proposed a new structure to synthesize the approximate optimal control by means of an artificial neural network. This neural optimal control is simultaneously used with two networks, one for the direct model of the controlled process and another for the controller, generating the control signal for the process. Artificial neural network together with long- and short-term memory method is adopted in the neural optimization control structure and the controller network expressed by the brain model with parameters. Almeida et al. pointed out the disadvantages of online training all the weights for the earlier neural networks: time consuming and slow convergence rate, easy to fall into the local optimal, and lacking intelligent use of a large number of data set. Using long- and short-term memory method, the weight is divided into two parts: one is related to long-term memory and the other is to short term. The former needs to be offline training with a large number of data sets so that the model can cover a large number of conditions and operating points. The latter needs to be online training, which can reflect the changes in instantaneous operating conditions. The shortcomings of earlier neural network methods can be overcome by this way [19].

Golbert et al. pointed out that the static gain is variable in the case of normal operation of the fuel cell or in the case of disturbance by analyzing the nonlinear model. Then they designed a multivariable nonlinear model predictor controller to adjust the power output. Golbert et al. also designed the predictive controller with order-decreased model to improve the fuel cell efficiency by selecting the appropriate objective function and constraints. However, their research is only based on simulation [20],[21].

Methekar et al. established a fuel cell model with distributed parameters and studied the dynamic characteristics of fuel cells by this model. They used the hydrogen and oxygen ratio control strategies to overcome the hypoxia in the cathode. They pointed out that it is difficult for the linear controller to keep up with the nonlinear dynamics due to the nonlinear characteristics of fuel cells. But what they did was also simulation research [22].

Yang and Wang considered the fuel cell system as a two-input, two-output system composed with hydrogen, air flow, output voltage, and current. They designed an adaptive controller on the basis of the identification of the low-order dynamic information of the system through the recursive least squares method and the introducing of the auxiliary items to represent the unmodeled and high-order dynamic information [23],[24].

Feroldi et al. implemented the dynamic matrix control method to realize the control of oxygen surplus rate and the stack voltage by manipulating the given voltage of the compressor and the control valve in the cathode outlet [25].

Danzer et al. established a dynamic model of the air supply system to achieve the multivariable control of the cathode pressure and the oxygen surplus rate by adjusting air flow and cathode outlet opening, which prevented hypoxia and improved overall efficiency. At the same time, Danzer et al. also designed the anode pressure observer [26].

Vega-Leal et al. presented a preliminary discussion on the temperature and air control [27] to achieve maximum overall efficiency and reduce the consumption of the compressor.

Over the past decade, the research on the control of PEM fuel cell has been very rich. Different methods, such as robust control [28], sliding mode control [29], model predictive control [30–31], and fuzzy PID control [32–33], have been utilized in the control of PEM fuel cell systems. The studies focus on not only the control of the air supply [34], but also the control of the oxygen excess ratio [35–36] and the tracking of the maximum efficiency point [37].

1.7 Problems in the fuel cell control

Fuel cell is a complex nonlinear, time-varying system. It is difficult to establish an accurate mathematical model. Taking into account the economy, reliability, and practicality, fuel cell devices often have different hardware structures and different adjustment means, which put forward higher requirements for the controller design as follows:

1) The PEM is the core component of the fuel cell stack, which determines the fuel cell life. In fact, it is a few tens of microns thick organic film. The flow of the air and hydrogen in the anode and cathode will cause a certain impact on the PEM. The anode and cathode pressure balance method is usually used to reduce the gas stress on the membrane. Although the general PID controller can meet the basic requirements of the balance of the anode and cathode pressure, it will cause fluctuation frequently in the steady state, making the force uneven in the PEM and reducing the life of the PEM, thereby causing the stack to age. For the advanced control methods, the algorithm is usually complex, which makes it difficult to run in real time due to the limitation of the controller hardware. It is necessary to develop a simple and practical control method to realize real-time control and precise control simultaneously, making the stress in the PEM as smooth as possible and extending the life of the stack.

2) In practical applications, fewer actuators will be used in terms of lowering the cost and reducing the weight, which often forms under-actuated systems that require the use of fewer operational variables to control more objects. Because the fuel cell is a complex nonlinear system, it is difficult to establish a precise mathematical model. These factors motivate the development or application of a model-free underactuated system control method to achieve the fuel cell control.

References

[1] Mao, Z.Q. 2005. Fuel Cell. Chemical Industry Press. Beijing, China.

[2] Xie, X.F., X.H Fan. 2004. Fuel Cell Technology. Chemical Industry Press. Beijing, China:Japanese Institute of Electrical Engineering. Special committee for the research of 21st century fuel cell power generation system technology.

[3] Zhang, Y. and B. Zhou. 2011. Modeling and control of a portable proton exchange membrane fuel cell – battery power system. Journal of Power Sources. 196: 8413–8423.

[4] Huang,Z., H.L. Tu, J.Q. Zhang and F. Zhan. 2000. Research, Development and Application of Proton Exchange Membrane Fuel Cell. Metallurgical Industry Press. Beijing, China.

[5] James, L., D. Andrew (Write), H. Zhu (Translate) and B.L. Yi (Check). 2006. Fuel Cell System – Principles, Design and Application. Science Press. Beijing, China.

[6] Larminie, J. and A. Dicks. 2013. Fuel Cell Systems Explained. John Wiley & Sons Ltd. West Sussex, England.

[7] Jurado, F. and J.R. Saenz. 2003. Adaptive control of a fuel cell-microturbine hybrid power plant. IEEE Transactions on Energy Conversion. 18: 342–347.

[8] Tsourapas,V., A.G. Stefanopoulou and J. Sun. 2007. Model-based control of an integrated fuel cell and fuel processor with exhaust heat recirculation. IEEE Transactions on Control Systems Technology. 15: 233–245.

[9] Kim, Y.H. and S.S. Kim. 1999. An electrical modeling and fuzzy logic control of a fuel cell generation system. IEEE Transactions on Energy Conversion. 14: 239–244.

[10] Pukrushpan, J.T., A.G. Stefanopoulou and H. Peng.2002. Modeling and control for PEM fuel cell stack system. In: Proceedings of the 2002 American Control Conference. Alaska, USA. 8–10 May 2002. IEEE. Piscataway, USA.
[11] Pukrushpan, J.T. 2003. Modeling and control of fuel cell systems and fuel processors. PhD Dissertation. University of Michigan. Michigan, USA.
[12] Suh, K.W. and A.G. Stefanopoulou. 2007. Performance limitations of air flow control in power-autonomous fuel cell system. IEEE Transactions on Control Systems Technology. 15: 465–473.
[13] Pukrushpan, J.T., A.G. Stefanopoulou and H. Peng. 2004. Control of fuel cell breathing. IEEE Control System Magazine. 24: 30–46.
[14] Rodatz, P., G. Paganelli and L. Guzzella. 2003. Optimizing air supply control of a PEM fuel cell system. In: Proceeding of the 2003 American Control Conference. Colorado, USA. 4–6 June 2003. IEEE. Piscataway, USA.
[15] Vahidi, A., A. Stefanopoulou and H. Peng. 2004. Model predictive control for starvation prevention in a hybrid fuel cell system. In: Proceeding of the 2004 American Control Conference. Massachusetts, USA. 30 June–2 July 2004. IEEE. Piscataway, USA.
[16] Vahidi, A., A. Stefanopoulou and H. Peng. 2006. Current management in a hybrid fuel cell power system: a model-predictive control approach. IEEE Transactions on Control Systems Technology. 14: 1047–1057.
[17] Caux, S., J. Lachaize, M. Fadel, P. shott and L. Nicod. 2005. Modeling and control of a fuel cell system and storage elements in transport applications. Journal of Process Control. 15: 481–491.
[18] Chen,G.L., X.H. Wen and L.W. Zhang. 2003. A coordinate double-loop PID pressure control method for PEMFC system. In: Proceedings of the Sixth International Conference on Electrical Machines and Systems. Beijing, China. 9–11 November 2003. Beijing World Publishing Corporation. Beijing, China.
[19] Almeida, P.E.M. and M.G. Simões. 2005. Neural optimal control of PEM fuel cell with parametric CMAC networks. IEEE Transactions on Industry Applications. 41: 237–245.
[20] Golbert, J. and D.R. Lewin. 2004. Model-based control of fuel cell: (1) Regulatory control. Journal of Power Sources. 135: 135–151.
[21] Golbert, J. and D.R. Lewin. 2007. Model-based control of fuel cell (2): Optimal efficiency. Journal of Power Sources. 173: 298–309.
[22] Methekar, R.N., V. Prasad and R.D. Gudi. 2007. Dynamic analysis and linear control strategies for proton exchange membrane fuel cell using a distributed parameter model. Journal of Power Sources. 165: 152–170.
[23] Yang, Y.P., F.C. Wang, H.P. Chang, Y.W. Ma and B.J. Weng. 2007. Low power proton exchange membrane fuel cell system identification and adaptive control. Journal of Power Sources. 164: 761–771.
[24] Wang, F.C., Y.P. Yang, C.W. Huang, H.P. Chang and H.T. Chen. 2007. System identification and robust control of a portable proton exchange membrane fuel-cell system. Journal of Power Sources. 164: 704–712.
[25] Feroldi, D., M. Serra and J. Riera. 2007. Performance improvement of a PEMFC system controlling the cathode outlet air flow. Journal of Power Sources. 169: 205–212.
[26] Danzer, M.A., J. Wilhelm, H. Aschemann, and E.P. Hofer. 2008. Model-based control of cathode pressure and oxygen excess ratio of a PEM fuel cell system. Journal of Power Sources.176: 515–522.
[27] Vega-Leal, A.P., F.R. Palomo, F. Barragán, C. García and J.J. Brey. 2007. Design of control systems for portable PEM fuel cells. Journal of Power Source. 169: 194–197.
[28] Wang, F. C., H. T. Chen, Y. P. Yang, et al. 2008. Multivariable robust control of a proton exchange membrane fuel cell system. Journal of Power Sources. 177: 393–403.

[29] Asai, Y. and N. Takahashi. 2010. Two-variable and two-degree-of-freedom sliding mode control for air pressure and flow in a fuel cell system. Journal of System Design & Dynamics. 4: 683–697.

[30] Gruber, J. K., C. Bordons and A. Oliva. 2012. Nonlinear MPC for the airflow in a PEM fuel cell using a Volterra series model. Control Engineering Practice. 20: 205–217.

[31] Schultze, M. and J. Horn. 2016. Modeling state estimation and nonlinear model predictive control of cathode exhaust gas mass flow for PEM fuel cells. Control Engineering Practice. 49:76–86.

[32] Ou, K., Y. X. Wang, Z. Z. Li, et al. 2015. Feedforward fuzzy-PID control for air flow regulation of PEM fuel cell system. International Journal of Hydrogen Energy. 40:11686–11695.

[33] Baroud, Z., M. Benmiloud, A. Benalia, et al. 2017. Novel hybrid fuzzy-PID control scheme for air supply in PEM fuel-cell-based systems. International Journal of Hydrogen Energy. 42: 10435–10447.

[34] Baroud, Z. M., Benmiloud and A. Benalia. 2015. Sliding mode controller for breathing subsystem on a PEM fuel cell system. In: International Conference on Control, Engineering & Information Technology. Tlemcen, Algeria. May 25–27 2015. IEEE. Piscataway, USA.

[35] Danzer, M. A. J. Wilhelm, H. Aschemann, et al. 2008. Model-based control of cathode pressure and oxygen excess ratio of a PEM fuel cell system. Journal of Power Sources. 176: 515–522.

[36] Wang, Y. X., and Y. B. Kim. 2014. Real-time control for air excess ratio of a PEM fuel cell system. IEEE/ASME Transactions on Mechatronics. 19: 852–861.

[37] Lu, J. and A. Zahedi. 2012. Air supply control for maximum efficiency point tracking in fuel cell systems. Journal of Renewable & Sustainable Energy. 4: 2430–2443.

2 Setup of the PEM fuel cell experiment platform

The proton exchange membrane (PEM) fuel cell experimental devices have different hardware architectures because of their different production processes, costs, and usage. With respect to the supply of raw materials, there are hydrogen-oxygen, hydrogen-air, direct methanol, direct ethanol, and direct diethyl ether PEM fuel cell systems; regarding the ways of humidification, there are internal humidification, external humidification, and self-humidification; with respect to the DC/DC designs, there are boost circuit and buck circuit, and many more. A 1 kW hydrogen-air PEM fuel cell experiment platform has been developed by us.

2.1 Overall architecture

The diagram in Figure 2.1 shows the 1 kW hydrogen-air PEM fuel cell experimental system [1]. For the sake of clarity and convenient study, the system is deliberately divided into six subsystems (see Figure 2.2), namely, gas supply subsystem, humidification subsystem, reaction subsystem, control subsystem, power conditioning subsystem, and load subsystems. The gas supply subsystem provides the fuel (hydrogen) and oxide (air) required for the reaction, which consists of a nitrogen tank, a hydrogen gas tank, an air compressor, an air filter, an air mass flow meter, and a hydrogen mass flow meter. Reaction subsystem is the core of the entire system. It is composed with a reaction stack, which is made up of 24 single-cell series. The activation area for each single cell is 160cm^2, the operating temperature is 65 °C, and the operating pressure is 30 kPa (gauge pressure). Hydrogen and air undergo a chemical reaction in the stack to produce electricity; at the same time, the by-products water and heat generated by the reaction are promptly removed. The output power of the reactor can reach 1 kW at appropriate temperature, humidity, and pressure. Humidification subsystem is used to increase the humidity of the hydrogen and air, by heating the circulating water, so as to improve the power generation efficiency. The goal of the control system is to follow the load changes by adjusting the hydrogen and air flow and to keep the system at the right temperature. The load subsystem is a power output unit. The power conditioning subsystem will output the voltage of the stack down to the standard 12 V DC voltage to provide direct current, and then provide 220 V AC through a DC/AC inverter. To meet the load rapid power change requirements, some auxiliary power supplies such as supercapacitor or battery are utilized to provide a stable power output for the load.

2.2 Gas supply subsystem

Considering the supply of raw materials, PEM fuel cells have two main categories: one is the hydrogen source fuel cell, including hydrogen-oxygen fuel cell and hydrogen-air

https://doi.org/10.1515/9783110602159-002

Figure 2.1: The PEM fuel cell experimental platform.

fuel cell. The other one uses methanol, ethanol, or other organic small molecules as fuel, namely, direct methanol fuel cell, direct ethanol fuel cells, and direct diethyl ether fuel cells. In the PEM fuel cells that use small organic molecules as fuel, the research of direct methanol fuel cell is the most extensive. Compared with hydrogen, methanol is a liquid at room temperature, is easy to store, has a high energy density, and has a wide range of sources. However, the strong methanol toxicity, anodic oxidation delay, and cathode performance attenuation caused by the methanol penetration are also great, which limits the development of methanol fuel cell. Other organic small molecule fuel cell researches can be divided into two categories based on whether C-C bond is used in molecular: one is the small organic molecule fuel cell with C-C bond and the other is the small organic molecule fuel cell free of C-C bond. Ethanol fuel cell, as an example of the former, has such characteristics that the electrocatalytic oxidation process is more complex; especially there is a problem that the C-C bond is hardly oxidized by the catalyst. There are two main questions for the C-C bond-free organic small molecule fuel cell [2]. First is the same fuel penetration problem as the direct methanol fuel cell because the intermediate product is methanol or formaldehyde, such as dimethoxy methane and trimethoxy methane. Second is regarding the low energy density and toxicity, such as formic acid and formaldehyde. In the process of reaction, fuel cells with small organic molecules as fuel produce carbon monoxide and carbon dioxide, which will cause damage to the environment. From the perspective of environmental protection, hydrogen energy fuel cell is the ideal choice, which is why most of the fuel cell city buses use hydrogen-based fuel cells. And from the economic point of view, hydrogen oxygen fuel cell is the most widely used. In this book, our researches focus on the hydrogen oxygen/air PEM fuel cell.

In the air-fueled fuel cell, there are usually two ways of air supply: one is controlling the air flow through regulation of the given voltage of the air compressor,

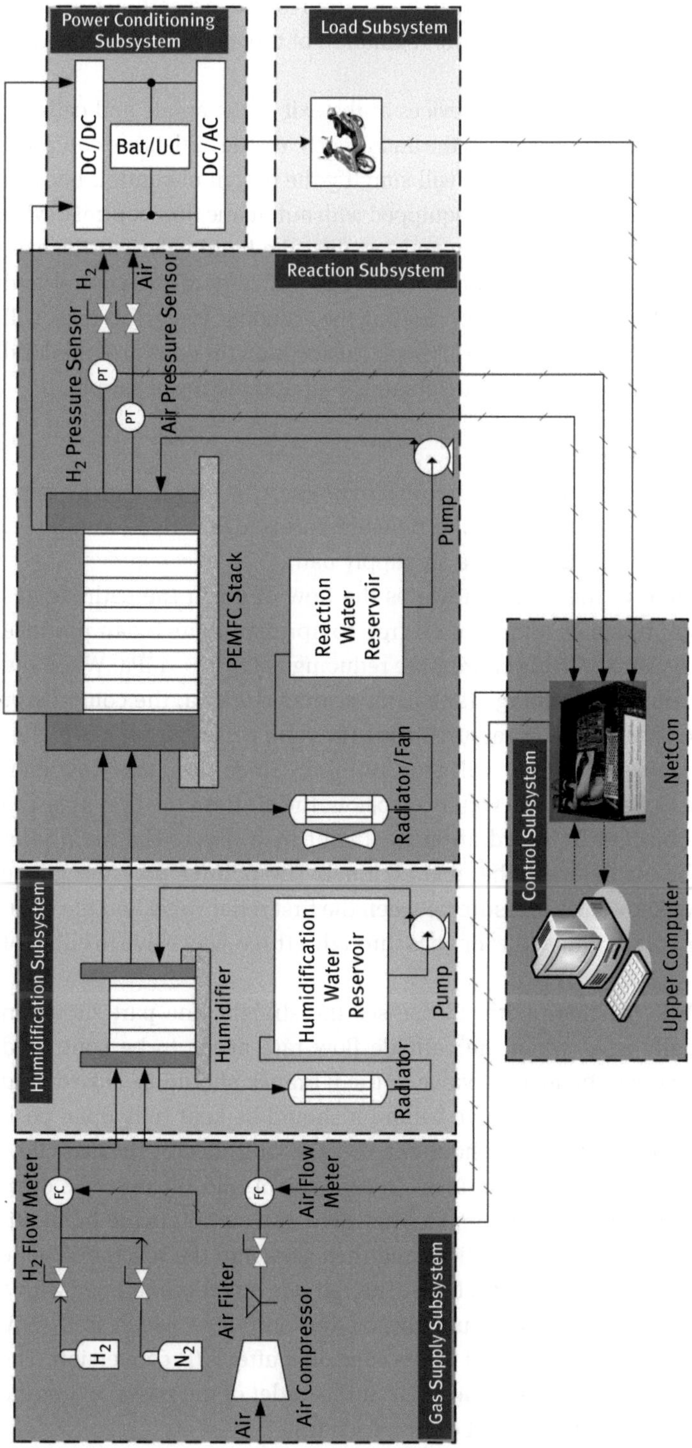

Figure 2.2: Structure diagram of the PEM fuel cell experiment platform.

and the other one is to use air compressors to boost air pressure to a certain pressure, and then control the air supply through the regulation of the air mass flowmeter. In this book, we adopt the second structure.

The design of the different fuel cell devices at the exit of the anode and cathode passages is also different. Some designs in the anode and cathode outlets are equipped with automatic flow control devices. This will simplify the design of control, but will increase the cost. Some of the designs are equipped with automatic flow control device at the entrance of the anode and cathode, and installed the hand valves at the exit, which greatly reduces the cost, but also increases the difficulty of the control and requires more advanced control means. Considering the economic issue, this book will adopt the latter structure. The research in this book will focus on the control of this kind of fuel cell device. The following details are about the air and hydrogen supply.

1) The design of the air supply channel

The air supply path is composed of an oil-free air compressor, an air filter, an air mass flow controller, a pressure sensor, a pointer type pressure gauge, valves, and pipes. Figure 2.3 is the structure diagram for the air supply path.

The thick solid line in the figure indicates the flow of air in the cathode gas passage. The air compressor is a kind of oil-free compressor, which can produce clean air. The setting value of the first pressure reducing valve is 0.2MPa. When the air pressure in the compressor storage tank is lower than 0.2 MPa, the compressor motor starts to work. When the air pressure inside the compressor air tank is higher than 0.7MPa, the compressor motor will stop running. Since the mass flow controller is susceptible to damage from water corrosion, the air from the first pressure reducing valve is subjected to dehydration treatment by a dryer. The main inlet valve can control the air entering the next channel. The pointer pressure gauge before the main valve shows the pressure between the first relief valve and the main valve. After the main intake valve, the air goes through a three-way valve to either of the two channels.

1) The first path is the main path. Please see the straight line part shown in Figure 2.3. It is usually used when the cathode flow rate needs to be controlled automatically. In this case, the manual valve in the B branch should be closed, and the manual pressure regulating the valve behind it should be kept fully open position. The air flow and pressure of the cathode channel are controlled by the mass flow controller. It is important to note that valves in channels A and B cannot be open simultaneously. Through the pressure regulating valve, dry air enters the humidification passage for sufficient humidification, and then goes into the reactor cathode channel involved in electrochemical reactions. The cathode outlet is equipped with a range of 0 ~ 100kPa piezoelectric pressure sensor, the output of which is 0 ~ 5 V analog voltage signal. This signal is sent to the controller after A/D conversion. The outlet valve can be used to regulate the air flow at the outlet of the passage, but the valve opening is generally kept constant during each test.

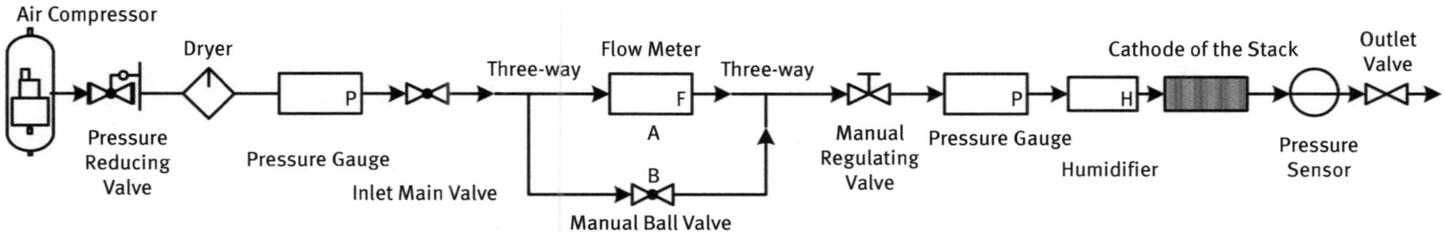

Figure 2.3: Structure diagram for the air supply path.

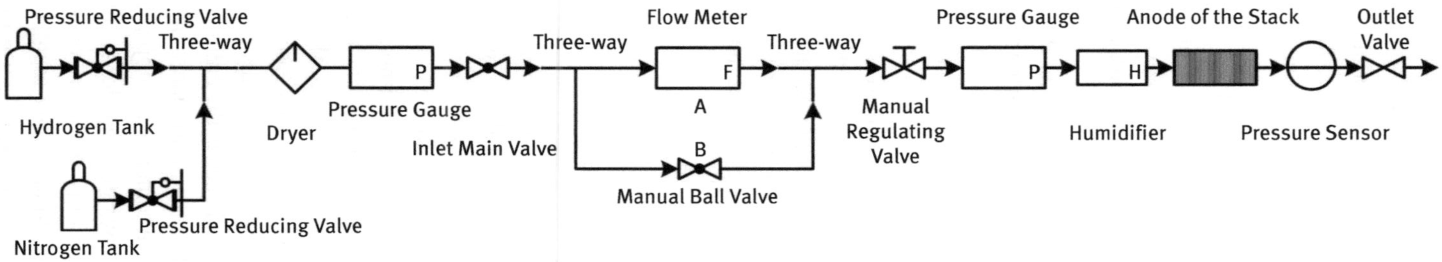

Figure 2.4: Structure diagram for the hydrogen supply path.

2) The other pathway is bypassed, usually used when not needing automatic control for the cathode flow rate, such as testing. In this case, the solenoid valve of the mass flow controller in the A channel should be closed, and the manual valve in B branch should be open. The flow and pressure of air in the cathode channel are adjusted by manually adjusting the subsequent pressure regulating valve.

2) The design of hydrogen supply channel

The hydrogen supply channel consists of a high pressure hydrogen tank, a high pressure nitrogen tank, a hydrogen mass flow controller, a pressure sensor, a pointer pressure gauge, and various valves and pipes. Figure 2.4 shows the structure of the hydrogen supply path.

The thick solid line in the figure indicates the flow of hydrogen gas in the anode gas channel. The design of the hydrogen supply path and the design of the air supply path are substantially similar. Two gas sources, hydrogen and nitrogen tanks, are used. Hydrogen is used for reactant and nitrogen is used for blowing treatment at the beginning and the end of every operation in order to ensure safety. Hydrogen and nitrogen tanks are equipped with manual ball valves, and only one valve can be open at a particular time. In addition, since the gas from the gas tank has been dried, dehydration is no longer needed. Therefore, no gas filter is installed in the anode gas line. An explosion-proof piezoelectric pressure sensor with a range of 0 ~ 100 kPa is installed at the anode outlet, the output of which is 4 ~ 20 mA current signal.

2.3 Humidification subsystem

In the PEM fuel cell operation, the proton conduction requires the presence of water. Decrease of the water content in the PEM will cause the conductivity of the membrane decrease linearly and then will lead to a decrease in cell performance. For an efficient operation, the stack needs to be maintained at a certain humidity. The effect of the humidification in the stack both takes an impact on the output performance of the stack and determines the stability and life of the reactor [3]. As the fuel cell stack is made up of several cells connected in series, the membrane resistance will increase when the film of a single cell in the stack loses of water and forfeits the ability to transfer protons. Then the produced ohmic heat will make the adjacent cells successively dehydrated and eventually cause the entire stack to fail [4]. On the other hand, too much water will also affect the performance of the stack. Stack generally supplies gas to all single cells through a common air supply channel. Each layer of gas flow rate may be uneven. When the flow rate of gas for a single cell in the stack is too small to bring out excess water, it will bring about the concentration polarization or even antipolar, leading the fuel and oxidants mix together, and then result in burning or explosion with the help of the catalyst to cause the stack to fail. Thus, the effective supply of water throughout the stack, as well as the uniformity of the water in the stack, becomes key to the design of the stack [5–6].

At present, the PEM fuel cell is mainly used for internal humidification, external humidification, self-humidification, etc [7].

1) Internal humidification

This method pumps the external liquid water into the internal battery, and then distributes the liquid water by the different internal structures, which can usually be achieved by designing the bipolar plate structure or designing the diffusion layer. Bipolar plate adopts a porous carbon plate inside the humidifier. The carbon plate has sufficient high porosity to enable the reactant gas to reach the catalytic layer smoothly. The water stored in the large number of microporous in the carbon plate can increase the humidity of the PEM. The water generated in the chemical reaction will be partially retained in the plate, maintaining the water storage capacity in the plate. Humidification technology for the diffusion layer uses a hydrophilic matrix layer and a hydrophobic substrate layer. The anode diffusion layer and the cathode diffusion layer all comprise a predetermined pore size of hydrophilic matrix layer. The base layer retains the wetting of the film by hydrophilic water absorption of the hydrophilic matrix layer.

2) External humidification

This is done by an additional device from outside. Generally, external humidification includes heating humidification, permeable membrane humidification, and direct liquid water injection humidification. Heating humidification is also known as bubbling humidification.

3) Self-humidification

This technology is realized by changing the internal structure of the battery. In contrast to internal humidification, self-humidification uses the water generated inside the battery to achieve humidification. Self-humidification technology can be used to simplify the stack system and improve the volume power ratio or weight ratio of the stack. But the amount of moisture is limited. It is only suitable for low-power, low-temperature type of stack.

As shown in Figure 2.5, our hydrogen-air PEM fuel cell experiment platform mainly uses the permeable membrane humidification of the external humidification mode. The heated deionized water in the humidification tank is pumped into the humidifier by the electromagnetic pump. The water flows from the top of the humidifier, along the strong hydrophilic humidifying membrane, down to the bottom of the humidifier, and then back to the tank. At the same time, air and hydrogen are also piped to the humidifier, along the other side of the humidification membrane. After being fully humidified through a special designed flow channel, the air and hydrogen flow from the humidifier to the reaction stack.

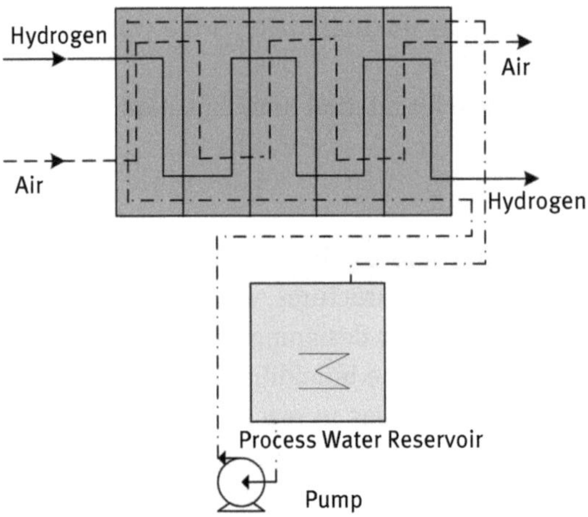

Figure 2.5: Schematic diagram of humidification system.

2.4 Power conditioning subsystem

The process of hydrogen PEM fuel cell is a nonlinear electrochemical reaction, which is a low-voltage, high-current device. The output voltage of the stack is about 0 ~ 25V. To provide a stable load current, the output voltage needs to be controlled by designing a DC/DC converter. One feasible design is the boost circuit, making it into a high-voltage, low-current device. Another design is the buck circuit. The output voltage of the former must be in accordance with the DC/AC standard voltage [8]. In order to use an existing DC/AC equipment, we designed a buck circuit. The stack output voltage is adjusted to a standard 12V DC voltage to output direct current, and then converted to 220V AC voltage through the DC/AC inverter to provide AC power.

The dynamic of fuel cell is relatively slow compared with some faster loads. There will be an instantaneous phenomenon that the fuel cell cannot keep up with the fluctuations of the load. In addition, the power of the load will exceed the maximum power that the fuel cell can provide in a short time. Therefore, auxiliary power is necessary to work together with fuel cells. These hybrid power supplies in general have three structures: the first structure is a hybrid system composed by fuel cell and secondary battery; the second structure is a hybrid system composed by fuel cell and supercapacitor; and the third structure is a hybrid system assembled by fuel cell, battery, and supercapacitor. In this book, the second structure is studied.

2.5 Control subsystem

Reliable control system is necessary for control of the flow, pressure, and other parameters of the fuel cell so as to keep it working properly and efficiently. dSPACE [9–10], which is a MATLAB/Simulink-based control system in real-time environment developed

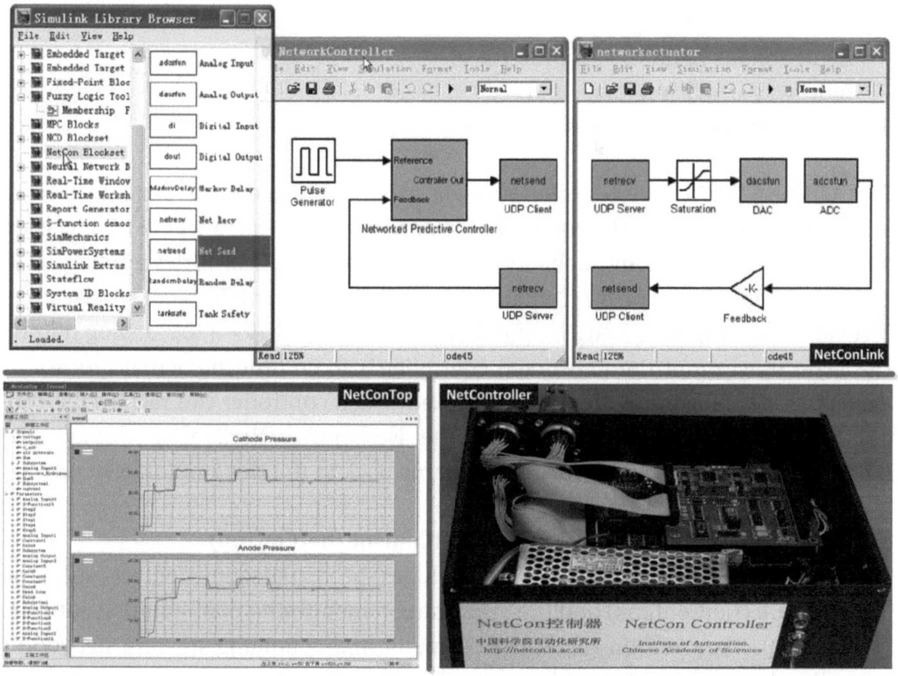

Figure 2.6: NetCon system.

by the dSPACE company in Germany, [9–10] or PLC (Programmable Logic Controller) is usually used in practice. The Institute of Automation, Chinese Academy of Sciences, has developed a MATLAB/Simulink-based networked control system (see Figure 2.6), which can be used as a controller for the fuel cell. The networked control system includes a hardware part – NetController – and a software part – NetConLink (visualization control configuration software) and NetConTop (visualization monitoring configuration software).

1) Networked control system hardware (NetController)

The NetController is the hardware part of the networked control system, as shown in Figure 2.6. It is the front-end execution unit of the networked control system, which is based on the 32-bit ARM9 embedded microprocessor. The operating frequency for the microprocessor can reach 200 to 500 MHz. The specific model is AT91RM9200 processor produced by ATMEL corporation. The microprocessor is based on the ARM920T core, designed for industrial control and communications [11]. Its main features are as follows:[/UIP]

- On-chip integrated ARM920T core, the operating frequency is 180 MHz;
- Embedded 16 KB data cache and 16KB instruction cache;
- 10 M/100 M Ethernet MAC interface;

- Full-speed main port and USB 2.0 full-speed second port;
- Synchronous Serial Control Interface;
- four- way UART interface; and
- Other peripheral interfaces.

The A/D and D/A converter interfaces are extended to meet the demand of data acquisition and control. The controller is intended with 16 channels 12-bit precision A/D converter interface and 2 channels 16-bit precision D/A converter interface. A/D converter chip adopts AD1674, analog signal input, and single/bipolar selectable; D/A converter chip uses DAC712, providing −5V ~ +5V, −10V ~ +10V analog output. The main function of the networked controller is to control the actual process by implementing the control algorithm. The networked controller can receive the control parameters and commands from the NetConLink/NetConTop through the Ethernet and feed the real-time running status back to the NetConLink/NetConTop.

2) Networked control system software (NetConLink/NetConTop)
The networked control system software includes the visualization control configuration software (NetConLink) and the visualization monitoring configuration software (NetConTop). They run on the upper computer and communicate with networked controller through an Ethernet interface. The system monitors the voltage, current, gas flow, pressure, and temperature of the reactor by means of cyclic monitoring and scanning. The software consists of the main program and a variety of subroutines. To facilitate program analysis, the software adopts structured module programming method. The modules are connected and called by some subroutines to achieve the required functions.

Control configuration software is based on Matlab/Simulink with Real-Time Workshop (RTW), which can achieve seamless integration with Simulink. Using various modules provided by Simulink and the S-functions for users, the models for controllers and processes can be established quickly, and the feasibility of the control strategy can be verified by the off-line and online simulation. Combined with RTW, code generation, compile, and link can be completed automatically within a few seconds, and then downloaded into the networked controller through the Ethernet interface on the controller to execute. As MATLAB/Simulink provides a block-diagram-based off-line simulation tool for the control system. The block diagrams for the control system can be easily and intuitively built through the visual interface and the drag-and-drop operation of the mouse. These block diagrams, which are simple and easy to modify, are the universal language for the control systems and control algorithms. Most system designs can be completed by calling the rich model library provided by the Simulink. If you need to add the control loops in the user's own design of the control algorithm or the hardware device interface module in the loop, you can define the user's own modules, which can be described by Simulink C-S functions. RTW is a toolbox for real-time control in Matlab and is often used with Simulink. It can translate the block diagram of Simulink into the portable C source code, and can generate the executable programs in different environments. RTW can analyze dynamic system models and support the vast majority of the computing platform. Through the RTW function, the user can quickly generate efficient C code for control system applications, and then implement in the Linux environment of the networked controller after translating the C code into an executable code [12].

Visualization monitoring configuration software is used to construct and generate graphical monitoring programs for real-time data acquisition and monitoring in the upper computers. It can realize the real-time and historical data processing, alarming, flow control, animation display, trend curve, and other functions for the fuel cell system.

The networked control system is a control system based on the network, which can be used as a network controller, a local controller, and a sensor. In this book, we use the networked control system to realize the control and monitoring of the PEM fuel cell system.

References

[1] Tong, S. W. and G. P. Liu. 2008. Real-time simplified variable domain fuzzy control of PEM fuel cell flow systems. European Journal of Control. 14: 223–233.

[2] Zhang, Q. 2007.Research on direct dimethyl ether fuel cell. Shandong University of Technology. Zibo, China.

[3] Deng, X. R. 2006.Modeling and control of proton exchange membrane fuel cell temperature and humidity system. Institute of Automation. Chinese Academy of Sciences. Beijing, China.

[4] Friede,W., S. Raël and B. Davat. 2004. Mathematical model and characterization of the transient behavior of a PEM fuel cell. IEEE Transactions on Power Electronics. 19: 1234–1241

[5] Chen, D. M. and H. Peng. 2004. Modeling and simulation of a PEM fuel cell humidification system. In: Proceeding of the 2004 American Control Conference. Massachusetts, USA. 30 June–2 July 2004. IEEE, Piscataway, USA.

[6] Chen, D. M. and H. Peng. 2005. Analysis of non-minimum phase behavior of PEM fuel cell membrane humidification systems. In: Proceeding of the 2005 American Control Conference. Portland, USA. 8–10 June 2005. IEEE. Piscataway, USA.

[7] James, L., D. Andrew(Write), H.Zhu (Translate) and B.L.Yi (Check),2006. Fuel Cell System – Principles, Design and Application.Science Press. Beijing, China.

[8] Liu, G. D. 2007. Research on the power management system of proton exchange membrane fuel cell. Institute of Automation, Chinese Academy of Sciences, Beijing, China.

[9] Jiang, Z. H., L. J. Gao, M. J. 2004. Blackwelder and R. A. Dougal, Design and experimental tests of control strategies for active hybrid fuel cell/battery power sources. Journal of Power Sources. 130: 163–171.

[10] Jiang, Z. H. and R. A. Dougal. 2004. Control strategies for active power sharing in a fuel-cell-powered battery-charging station. IEEE Transactions on Industry Applications. 40: 917–924.

[11] ATMEL. 2000. AT91RM9200 User Mannal, www.atmel.com/dyn/products.

[12] Xiong, Y. F. 2006. Research on the modeling and control of a small proton exchange membrane fuel cell flow system. Institute of Automation, Chinese Academy of Sciences. Beijing, China.

3 Modeling of the PEM fuel cell system

The hydrogen-air proton exchange membrane (PEM) fuel cell is inherently a high-order, nonlinear system. Besides the electrochemical reaction taken in the stack, it also involves fluid flow, heat transfer, and energy conversion processes. Many literature has already built all kinds of mechanism models of the fuel cell. For convenient controller design and validation, practice-oriented models are set up in this book. On the one hand, some of the parameters in the mechanism model are from the actual experimental platform. On the other hand, some subsystem models such as the gas flow subsystem are directly obtained by identification. Furthermore, Simulink models for the 1 kW hydrogen-air PEM fuel cell is also established since Simulink provides us many powerful tools for analysis and design of a control system.

3.1 Stack voltage model

The fuel cell stack usually consists of N cells in series. The stack voltage v_{st} can be calculated by $v_{st} = N \times v_{cell}$ where v_{cell} is the voltage of each cell. The document [1] gives the basic expression of the cell voltage:

$$v_{cell} = E - v_{act} - v_{ohm} - v_{con} \tag{3.1}$$

where the first term E on the right side of the equation is the reversible voltage. It can be calculated by the following formula: [2–3]

$$E = 1.229 - \left(8.5 \times 10^{-4}\right)\left(T_{st} - 298.15\right) + 4.308 \times 10^{-5}\, T_{st}\left(\ln\left(P_{H_2}\right) + 0.5\ln\left(P_{O_2}\right)\right) \tag{3.2}$$

The second term v_{act} is the activation voltage loss which is used to cause the electrochemical reaction to proceed toward the direction of water generation and electricity, rather than in the opposite direction. The activation voltage is determined by the following equation:

$$v_{act} = a_1 + a_2 T_{st} + a_3 T_{st} \ln(i_{st}) + a_4 T_{st} \ln\left(c_{O_2}\right) \tag{3.3}$$

where a_1, a_2, a_3, and a_4 are the parameters to be identified,

$$c_{O_2} = P_{O_2} / \left(5.08 \times 10^6 \exp(-498/T_{st})\right)$$

The third term v_{ohm} is the ohmic voltage loss, which is similar to the voltage loss in a conventional electronic device in order to overcome the protons passing through the solid electrolyte and the electrons on its conductive path. The ohmic voltage loss is calculated by

https://doi.org/10.1515/9783110602159-003

$$v_{ohm} = i_{st} \cdot R_{ohm} = i_{st} \cdot \left(R_{pro} + R_{ele}\right) \tag{3.4}$$

where R_{ele} is the resistance of the electron flow and can be regarded as a constant. R_{pro} is the resistance of the proton flow, which is determined by the following equation:

$$R_{pro} = r_m \cdot \sigma_m / A$$

where r_m represents the specific impedance of the exchange membrane to the humidified proton flow. For Nafion 112 (51 μm) film, the empirical formula is:

$$r_m = \frac{181.6\left(1 + 0.03\left(\frac{i}{A}\right) + 0.062\left(\frac{T}{303}\right)^2\left(\frac{i}{A}\right)^{2.5}\right)}{\left(\lambda - 0.634 - 3\left(\frac{i}{A}\right)\right)\exp\left(4.18\left(\frac{T-303}{T}\right)\right)}$$

The last term v_{con} represents the voltage loss due to the decrease in the concentration of the reaction gas and the mass transfer of oxygen and hydrogen.

$$v_{con} = B \cdot \ln\left(1 - \frac{i}{i_{max}}\right) \tag{3.5}$$

In our modeling, the bipolar plate capacitance phenomenon reflecting the performance of the fuel cell is also considered [4].

$$\frac{d(v_{act} + v_{con})}{dt} = \frac{i_{st}}{C_{double}} - \frac{v_{act} + v_{con}}{(R_{act} + R_{con})C_{double}} \tag{3.6}$$

where $R_{act} = v_{act}/i_{st}$, $R_{con} = v_{con}/i_{st}$.

3.2 Cathode flow model

The voltage of the fuel cell is mainly affected by the partial pressure of the gas as shown by equations (3.2) and (3.3). Therefore, the fuel cell cathode flow model should fully take into account the gas partial pressure. By applying the law of mass conservation, we can get equations (3.7), (3.8), and (3.9).

$$\frac{dP_{O_2,ca}}{dt} = \frac{\mathcal{R}T_{ca}}{V_{ca}}\left(F_{O_2,ca,in} - F_{O_2,rea} - F_{O_2,ca,out}\right) \tag{3.7}$$

$$\frac{dP_{N_2,ca}}{dt} = \frac{\mathcal{R}T_{ca}}{V_{ca}}\left(F_{N_2,ca,in} - F_{N_2,ca,out}\right) \tag{3.8}$$

$$\frac{dP_{H_2O,ca}}{dt} = \frac{\mathcal{R}T_{ca}}{V_{ca}}\left(F_{H_2O,ca,in} - k_{vap}F_{H_2O,gen} - F_{H_2O,ca,out}\right) \tag{3.9}$$

In a hydrogen-fueled fuel cell system, the input gas is dry air, including 21% oxygen and 79% nitrogen.

$$F_{O_2,ca,in} = f_{1,O_2} \cdot F_{air,in} = 0.21F_{air,in}$$

$$F_{N_2,ca,in} = f_{1,N_2} \cdot F_{air,in} = 0.79F_{air,in}$$

Dry air will be humidified before entering the cathode path. Thus, the vapor carried by the air into the cathode can be calculated by the following formula:

$$F_{H_2O,ca,in} = f_{1,H_2O} \cdot F_{air,in} = \frac{P_{H_2O,ca}}{P_{ca} - P_{H_2O,ca}} F_{air,in}$$

From the thermodynamic knowledge of the mixed gas, the partial pressure of the vapor in the cathode is

$$P_{H_2O,ca} = \varnothing_{ca} \cdot P_{sat}(T_{ca})$$

According to the principle of electrochemistry, the consumed oxygen and the produced water can be calculated by

$$F_{O_2,rea} = \frac{Ni_{st}}{4\mathcal{F}}$$

$$F_{H_2O,gen} = \frac{Ni_{st}}{2\mathcal{F}}$$

The volume flow rate of oxygen, nitrogen, and vapor at the cathode inlet can be calculated as the molar fraction of the total gas flow rate.

$$F_{O_2,ca,out} = f_{2,O_2} \cdot F_{ca,out} = \frac{P_{O_2,ca,out}}{P_{ca,out}} F_{ca,out}$$

$$F_{N_2,ca,out} = f_{2,N_2} \cdot F_{ca,out} = \frac{P_{N_2,ca,out}}{P_{ca,out}} F_{ca,out}$$

$$F_{H_2O,ca,out} = f_{2,H_2O} \cdot F_{ca,out} = \frac{P_{H_2O,ca,out}}{P_{ca,out}} F_{ca,out}$$

According to the principle of mass conservation, the volume flow rate at the cathode outlet has the following relationship:

$$F_{ca,out} = F_{O_2,ca,out} + F_{N_2,ca,out} + F_{H_2O,ca,out}$$

Applying Dalton's law of partial pressure

$$P_{ca} = P_{O_2,ca} + P_{N_2,ca} + P_{H_2O,ca}$$

Since the volume of the cathode path is very small and the distance from the cathode outlet valve to the stack is relatively close, the cathode pressure can be considered equal to the pressure at the cathode outlet, that is, $P_{ca,\,out} = P_{ca}$.

The mass flow rate of the gas at the cathode outlet can be described by the following nonlinear equation:

$$W_{ca,out} = a \cdot \left(-1 + \sqrt{1 + b \cdot (P_{ca} - P_{atm})} \right)$$

where a, b are the parameters to be identified. The mass flow rate can be converted to the volume flow rate by the equation $F_{ca,out} = k_{wf} W_{ca,out}$

In summary, the following partial pressure equations are obtained:

$$\frac{dP_{O_2,ca}}{dt} = \frac{\mathcal{R}T_{ca}}{V_{ca}} \left(0.21 F_{air,in} - \frac{Ni_{st}}{4\mathcal{F}} - \frac{P_{O_2,ca}}{P_{ca}} F_{ca,out} \right) \tag{3.10}$$

$$\frac{dP_{N_2,ca}}{dt} = \frac{\mathcal{R}T_{ca}}{V_{ca}} \left(0.79 F_{air,in} - \frac{P_{N_2,ca}}{P_{ca}} F_{ca,out} \right) \tag{3.11}$$

$$\frac{dP_{H_2O,ca}}{dt} = \frac{\mathcal{R}T_{ca}}{V_{ca}} \left(\frac{\varnothing_{ca} \cdot P_{sat}(T_{ca})}{P_{ca} - \varnothing_{ca} \cdot P_{sat}(T_{ca})} F_{air,in} - k_{vap} \frac{Ni_{st}}{2\mathcal{F}} - \frac{P_{H_2O,ca}}{P_{ca}} F_{ca,out} \right) \tag{3.12}$$

From equations (3.10), (3.11), and (3.12), the cathode pressure equation is obtained by applying Dalton's law of partial pressure:

$$\begin{aligned} \frac{dP_{ca}}{dt} = \frac{k_{wf}\mathcal{R}T_{ca}}{V_{ca}} \Bigg\{ & \frac{\varnothing_{ca} \cdot P_{sat}(T_{ca})}{P_{ca} - \varnothing_{ca} \cdot P_{sat}(T_{ca})} W_{air,in} - \frac{Ni_{st}}{4\mathcal{F}k_{wf}} \\ & + k_{vap} \frac{Ni_{st}}{2\mathcal{F}k_{wf}} - a \cdot \left(-1 + \sqrt{b \cdot (P_{ca} - P_{atm})} \right) \Bigg\} \end{aligned} \tag{3.13}$$

3.3 Anode flow model

Similar to the derivation of the cathode flow model, the anode flow model can be obtained. Here the anode gas partial pressure equation is:

$$\frac{dP_{H_2,an}}{dt} = \frac{\mathcal{R}T_{an}}{V_{an}} \left(F_{H_2,an,in} - \frac{Ni_{st}}{2\mathcal{F}} - \frac{P_{H_2,an,out}}{P_{an,out}} F_{an,out} \right) \tag{3.14}$$

$$\frac{dP_{H_2O,an}}{dt} = \frac{\mathcal{R}T_{an}}{V_{an}} \left(\frac{\varnothing_{an} \cdot P_{sat}(T_{an})}{P_{an} - \varnothing_{an} \cdot P_{sat}(T_{an})} F_{H_2,in} - \frac{P_{H_2O,an,out}}{P_{an,out}} F_{an,out} \right) \tag{3.15}$$

and the anode pressure equation is given directly.

$$\begin{aligned} \frac{dP_{an}}{dt} = \frac{k'_{wf}\mathcal{R}T_{an}}{V_{an}} \Bigg\{ & \frac{P_{an}}{P_{an} - \varnothing_{an} \cdot P_{sat}(T_{an})} W_{H_{2},in} - \frac{Ni_{st}}{2\mathcal{F}k'_{wf}} \\ & -c \cdot \left(-1 + \sqrt{1 + d \cdot (P_{an} - P_{atm})} \right) \Bigg\} \end{aligned} \tag{3.16}$$

3.4 DC/DC converter model

The DC/DC converter shown in Figure 3.1 is a buck circuit that reduces the DC voltage output from the stack to 12 V. It consists of a MOSFET tube, an inductor, a diode, and a capacitor. The DC/DC converter voltage can be achieved by adjusting the duty cycle of the Pulse Width Modulation (PWM) wave.

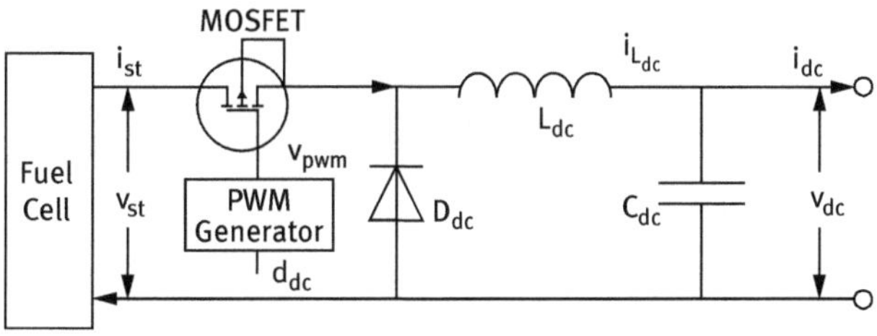

Figure 3.1: DC/DC circuit.

When $\upsilon PWM = 1$, the diode turns off.

$$\begin{cases} \dfrac{di_{L_{dc}}}{dt} = \dfrac{1}{L_{dc}} v_{st} - \dfrac{1}{L_{dc}} v_{dc} \\ \dfrac{dv_{dc}}{dt} = \dfrac{1}{C_{dc}} i_{L_{dc}} - \dfrac{1}{C_{dc}} i_{dc} \end{cases} \tag{3.17}$$

When $\upsilon PWM = 0$, the MOSFET is closed. Thus, it forms a closed loop between the diode, the inductor, and the capacitor.

$$\begin{cases} \dfrac{di_{L_{dc}}}{dt} = -\dfrac{1}{L_{dc}} v_{dc} \\ \dfrac{dv_{dc}}{dt} = \dfrac{1}{C_{dc}} i_{L_{dc}} - \dfrac{1}{C_{dc}} i_{dc} \end{cases} \tag{3.18}$$

According to equations (3.17) and (3.18), the output of any switch on or off can be obtained using the boundary condition. However, the switch has to actually act dozens of times or even hundreds of times to complete an adjustment process. Therefore the calculation is very heavy. In fact, the two equations can be combined into an equation using a weighted average method [5].

$$\begin{cases} \dfrac{di_{L_{dc}}}{dt} = \dfrac{d_{dc}}{L_{dc}} v_{st} - \dfrac{1}{L_{dc}} v_{dc} \\ \dfrac{d_{v_{dc}}}{dt} = \dfrac{1}{C_{dc}} i_{L_{dc}} - \dfrac{1}{C_{dc}} i_{dc} \end{cases} \tag{3.19}$$

3.5 Simulink model

Matlab/Simulink is a powerful tool for mathematical modeling and control system design. It is the international first selective computer language in the area of automatic control.

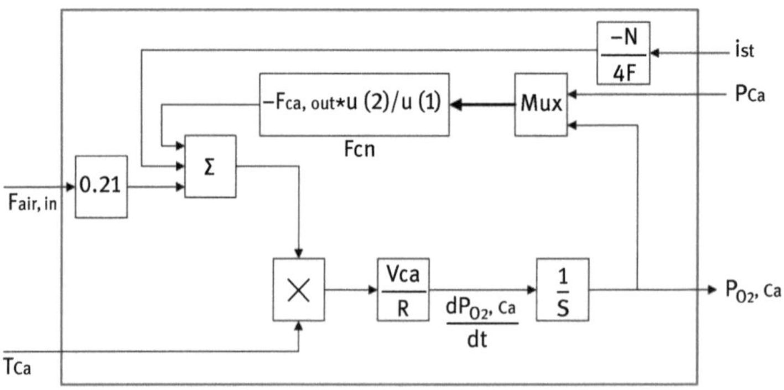

Figure 3.2: Oxygen partial pressure model structure.

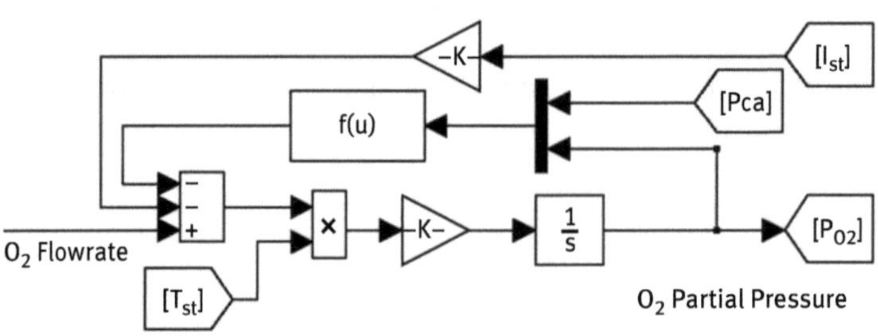

Figure 3.3: Oxygen partial pressure Simulink model.

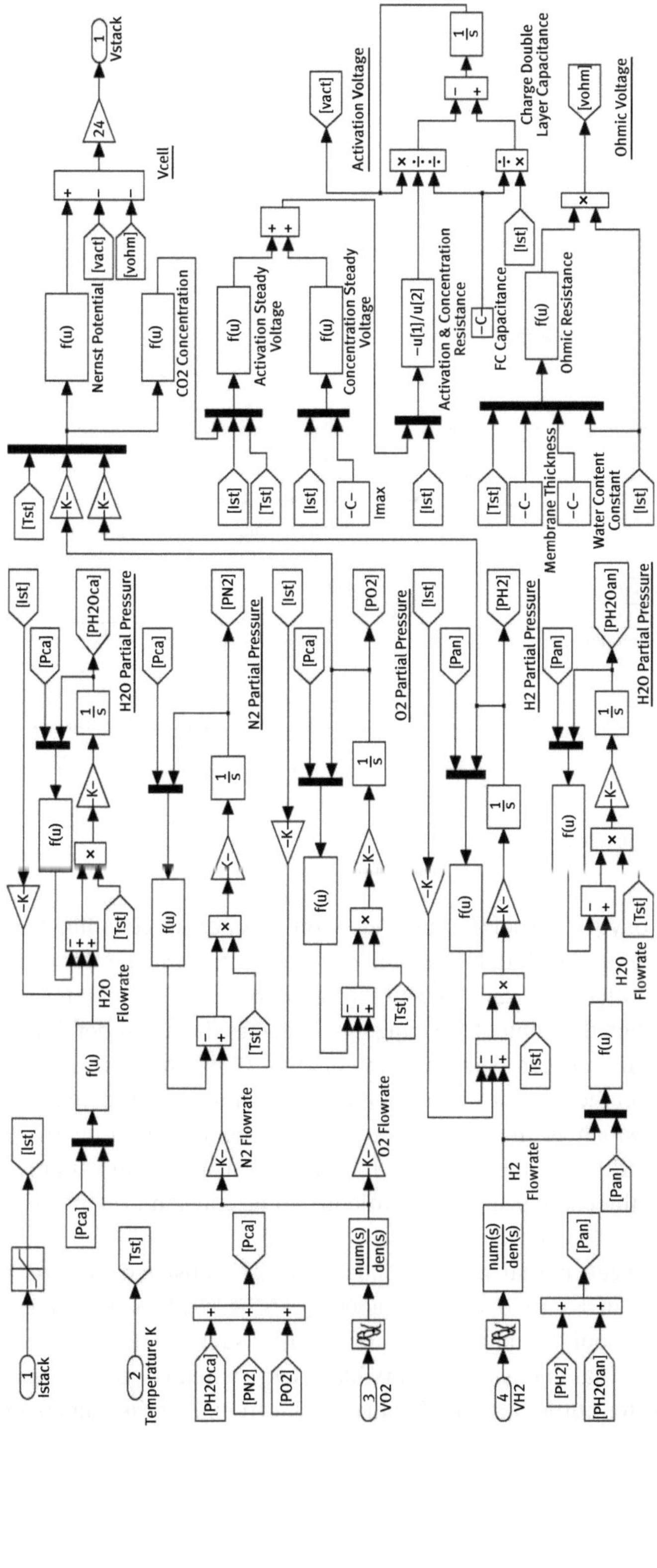

Figure 3.4: Hydrogen-air proton exchange membrane fuel cell model.

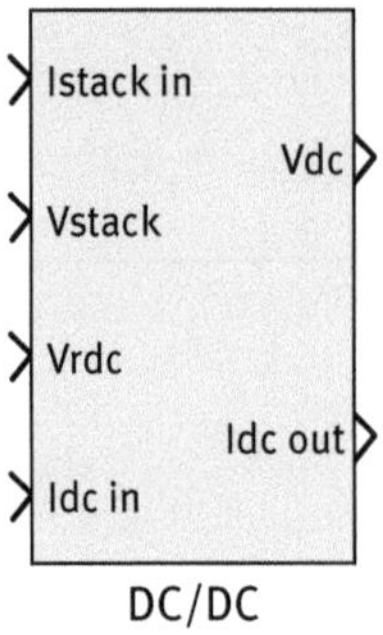

Figure 3.5: DC/DC module interface.

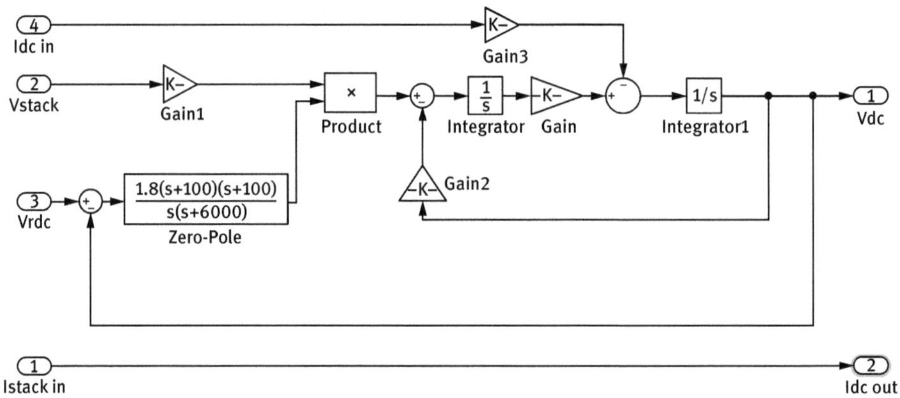

Figure 3.6: DC/DC Simulink model.

Let us take the oxygen partial pressure equation as an example to illustrate how to establish a model in the Matlab/Simulink environment. An integrator can be introduced for each differential variable so that the output of the integrator is the state variable itself. Then the input of the integrator becomes the first order differentiation of the state variable. It can be seen from equation (3.10) that the state variable is $P_{O_2,ca}$. The item on the left side of the equation is the first order differentiation of the state variable, which is the summation of three items. Selecting the variables T_{ca}, $F_{air,in}$, i_{st}, and P_{ca} as the inputs of the model and the state variable $P_{O_2,ca}$ as the output, and then supposing other variables to be constant, the model structure can be described by Figure 3.2.

Then the oxygen partial pressure model can be translated into the Simulink model as shown in Figure 3.3. Other models can be built in the same way (see Figure 3.4) [6]. It should be noted that many modules such as Mux or Demux support vector operation. That is to say, the Mux block can combines several input signals into a signal vector output and the Demux block can extract the components of an input signal and outputs the components as separate signals. The Fcn module is used to describe the mathematical operation of the input signal, where the input signal is the state vector of the system and the input signal in the Fcn module is denoted by u, and if u is a vector, u[i] stands for the ith variable. In order to make the model more concise and clear, the transfer modules Goto and From are utilized in the modeling process.

The DC/DC model is relatively complex due to the voltage that should be controlled. Figure 3.5 shows the input and output interfaces of the DC/DC module in Simulink. The voltage controller can be designed by zero-pole assignment. Figure 3.6 displays the detailed model of the DC/DC converter.

References

[1] Pukrushpan, J.T., A.G. Stefanopoulou and H. Peng.2002. Modeling and control for PEM fuel cell stack system. In: Proceedings of the 2002 American Control Conference. Alaska, USA. 8–10 May 2002. IEEE. Piscataway, USA.

[2] Khan, M. J. and M. T. Iqbal. 2005. Dynamic modeling and simulation of a small wind–fuel cell hybrid energy system. Renewable Energy. 30: 421–439.

[3] Mann, R. F., J. C. Amphlett, M. A. I. Hooper, H. M. Jensen, B. A. Peppley and P. R. Roberge. 2000. Development and application of a generalised steady-state electrochemical model for a PEM fuel cell. Journal of Power Sources. 86: 173–180.

[4] Xiong, Y. F. 2006. Research on the modeling and control of a small proton exchange membrane fuel cell flow system. Institute of Automation, Chinese Academy of Sciences. Beijing, China.

[5] Liu, G. D. 2007. Research on the power management system of proton exchange membrane fuel cell. Institute of Automation, Chinese Academy of Sciences, Beijing, China.

[6] Tong, S. W., D. W. Qian, J. J. Fang and H. X. Li. 2013. Integrated modeling and variable universe fuzzy control of a hydrogen-air fuel cell system. International Journal of Electrochemical Science. 8: 3636–3652.

4 Control of PEM fuel cell without load current feedback

Different control strategies could be taken to meet the different requirements of the working environment of fuel cells. If the fuel cell works at the optimum efficiency point, hydrogen consumption will be the lowest. The control strategy is relatively simple at this moment, and only needs to maintain the pressure balance between anode and cathode as much as possible, regarding the load as a disturbance. In the controller design, the cathode channel is considered as the main control loop, and the anode pressure follows the cathode pressure changes. Taking into account the real-time operation, control algorithm should be designed as simple as possible. In this chapter, a linear control method, the adaptive state feedback predictive control (ASFPC) method based on Singular Pencil Model (SPM) [1], is first proposed. The method jointly implements the state and parameter information of the controlled process to improve the control performance. Then, a nonlinear control method, the real-time simplified variable domain fuzzy control (VDFLC) [2], is presented, which can realize the real-time control and the precise control simultaneously.

4.1 SPM-based ASFPC

4.1.1 Joint estimation of states and parameters based on SPM

The multi-input multi-output (MIMO) linear dynamic system can be described with auto-regressive and moving average model.

$$\begin{aligned}(I+\tilde{A}_1 z^{-1}+\cdots+\tilde{A}_n z^{-n})y_k &= (\tilde{B}_1 z^{-1}+\cdots+\tilde{B}_m z^{-m})u_k \\ &\quad + (I+\tilde{C}_1 z^{-1}+\cdots+\tilde{C}_n z^{-n})e_k\end{aligned} \tag{4.1}$$

where y_k is the output vector, u_k is the input vector, e_k is the Gaussian white noise with mean 0, and z is the forward shift operator.

$$\underset{(n\times p)}{\tilde{A}_i} = \begin{bmatrix} a_{1;n_1-i,1} & \cdots & a_{1;n_1-i,p} \\ \vdots & \ddots & \vdots \\ a_{n;n-i,1} & \cdots & a_{n;n-i,p} \end{bmatrix}$$

$$\underset{(n\times m)}{\tilde{B}_i} = \begin{bmatrix} b_{1;n_1-i,1} & \cdots & b_{1;n_1-i,m} \\ \vdots & \ddots & \vdots \\ b_{n;n-i,1} & \cdots & b_{n;n-i,m} \end{bmatrix}$$

https://doi.org/10.1515/9783110602159-004

$$\underset{(n\times p)}{\tilde{C}_i} = \begin{bmatrix} c_{1;n_1-i,1} & \cdots & c_{1;n_1-i,p} \\ \vdots & \ddots & \vdots \\ c_{n;n-i,1} & \cdots & c_{n;n-i,p} \end{bmatrix}$$

1) System description with SPM

SPM model is a special form of expression [3, 4], which can be used to describe the linear dynamic system as follows:

$$[E - FD \quad G]\begin{bmatrix} x \\ w \end{bmatrix} = 0 \tag{4.2}$$

where $x \in \mathbb{R}^n$ is a state variable, also called an auxiliary variable or an internal variable; $w \in \mathbb{R}^{p+m}$ is an external variable consisting of the input and output variables of the system; p and m are known; E and F are $n+p$ by n matrices; D is a linear operator; G is a $n+p$ by $p+m$ matrix; and D can be a Laplace operator s for a continuous system and a forward shift operator z for a discrete system.

The state space model of the MIMO dynamic system can be expressed by the SPM model:

$$\begin{bmatrix} A - Iz & 0 & B \\ C & -I & D \end{bmatrix} \begin{bmatrix} x_k \\ y_k \\ u_k \end{bmatrix} \tag{4.3}$$

where u_k is the input vector, y_k is the output vector, and (A, B, C, D) is the state space matrix.

Auto-regressive moving average model

$$\sum\nolimits_{i=0}^{n} A_i z^{n-i} y_k = \sum\nolimits_{i=0}^{m} B_i z^{m-i} u_k \tag{4.4}$$

can also be expressed by SPM model:

$$\left[\begin{array}{cccc|c|c} -Iz & & & & -A_n & B_n \\ I & \ddots & & & \vdots & \vdots \\ & \ddots & -Iz & & \vdots & \vdots \\ & & I & & -A_0 & B_0 \end{array}\right] \begin{bmatrix} x_k \\ y_k \\ u_k \end{bmatrix} = 0 \tag{4.5}$$

SPM model has been widely used because it has the following characteristics [5–11]:

- Uniform: SPM model can not only represent the state space model, but also can represent the input-output models, and these models are the most commonly used models in linear control system.
- Simple: The matrix transformation can be done by performing elementary row or column transformations on the composition matrix E, F, and G of equation (4.2).
- General: Unlike the state space model, the SPM model can represent some of the concepts that can be useful in control system design, such as improper systems, PID controllers, and inverse of strictly proper systems.

The linear dynamic system represented by equation (4.1) can be described by the SPM model as:

$$\left[\begin{array}{c:c:c:c} E_* - Iz & -A_* & B_* & C_* \\ \hline E_0 & -I & 0 & I \end{array}\right]\begin{bmatrix} x_k \\ y_k \\ u_k \\ e_k \end{bmatrix} = 0 \tag{4.6}$$

where

$$\underset{(n\times p)}{A_*} = \begin{bmatrix} A_1 \\ \vdots \\ A_p \end{bmatrix}, \quad \underset{(n_i\times p)}{A_i} = \begin{bmatrix} a_{i;0,1} & \cdots & a_{i;0,p} \\ \vdots & \ddots & \vdots \\ a_{i;n_i-1,1} & \cdots & a_{i;n_i-1,p} \end{bmatrix}, \quad \underset{(n\times n)}{E_*} = \begin{bmatrix} E_1 & & \\ & \ddots & \\ & & E_p \end{bmatrix},$$

$$\underset{(n\times m)}{B_*} = \begin{bmatrix} B_1 \\ \vdots \\ B_p \end{bmatrix}, \quad \underset{(n_i\times m)}{B_i} = \begin{bmatrix} b_{i;0,1} & \cdots & b_{i;0,m} \\ \vdots & \ddots & \vdots \\ b_{i;n_i-1,1} & \cdots & b_{i;n_i-1,m} \end{bmatrix}, \quad \underset{(n_i\times n_i)}{E_i} = \begin{bmatrix} 0 & \cdots & 0 & 0 \\ 1 & \cdots & 0 & 0 \\ \vdots & \ddots & \vdots & \vdots \\ 0 & \cdots & 1 & 0 \end{bmatrix},$$

$$\underset{(n\times p)}{C_*} = \begin{bmatrix} C_1 \\ \vdots \\ C_p \end{bmatrix}, \quad \underset{(n_i\times p)}{C_i} = \begin{bmatrix} c_{i;0,1} & \cdots & c_{i;0,p} \\ \vdots & \ddots & \vdots \\ c_{i;n_i-1,1} & \cdots & c_{i;n_i-1,p} \end{bmatrix},$$

$$\underset{(p\times n)}{E_0} = \begin{bmatrix} 0 & \cdots & \overset{n_1^*}{1} & \cdots & \overset{n_i^*}{0} & \cdots & \overset{n_p^*}{0} \\ 0 & \cdots & 0 & \cdots & 1 & \cdots & 0 \\ & & \cdots & \cdots & \cdots & & \\ 0 & \cdots & 0 & \cdots & 0 & \cdots & 1 \end{bmatrix}, \quad \begin{array}{l} n_i^* = \sum_{j=1}^{i} n_j \\ n = n_p^* \end{array}$$

The following two examples illustrate how to use the SPM model to represent a linear system.
Example 4.1: Consider the following two-input, two-output system:

$$
\begin{aligned}
&\left\{\begin{bmatrix} 1 & 0 \\ 0 & 1 \end{bmatrix} + \begin{bmatrix} a_{111} & a_{112} \\ a_{211} & a_{212} \end{bmatrix} z^{-1} + \begin{bmatrix} a_{121} & a_{122} \\ a_{221} & a_{222} \end{bmatrix} z^{-2} \right\} y_k \\
&= \left\{ \begin{bmatrix} b_{111} & b_{112} \\ b_{211} & b_{212} \end{bmatrix} z^{-1} + \begin{bmatrix} b_{121} & b_{122} \\ b_{221} & b_{222} \end{bmatrix} z^{-2} \right\} u_k \\
&+ \left\{ \begin{bmatrix} 1 & 0 \\ 0 & 1 \end{bmatrix} + \begin{bmatrix} c_{111} & c_{112} \\ c_{211} & c_{222} \end{bmatrix} z^{-1} + \begin{bmatrix} c_{121} & c_{122} \\ c_{221} & c_{222} \end{bmatrix} z^{-2} \right\} e_k
\end{aligned}
$$

which can be described by SPM model with the following form:

$$
\left[\begin{array}{cccc|cc|cc|cc}
-z & 0 & 0 & 0 & -a_{121} & -a_{122} & b_{121} & b_{122} & c_{121} & c_{122} \\
1 & -z & 0 & 0 & -a_{111} & -a_{112} & b_{111} & b_{112} & c_{111} & c_{112} \\
0 & 0 & -z & 0 & -a_{221} & -a_{222} & b_{221} & b_{222} & c_{221} & c_{222} \\
0 & 0 & 1 & -z & -a_{211} & -a_{212} & b_{211} & b_{212} & c_{211} & c_{212} \\
\hline
0 & 1 & 0 & 0 & -1 & 0 & 0 & 0 & 1 & 0 \\
0 & 0 & 0 & 1 & 0 & -1 & 0 & 0 & 0 & 1
\end{array}\right]
\begin{bmatrix} x_k \\ y_k \\ u_k \\ e_k \end{bmatrix} = 0
$$

Example 4.2: Consider the following single-input, single-output system:

$$
\begin{aligned}
(1 + a_1 z^{-1} + a_2 z^{-2} + a_3 z^{-3})\ y_k &= (b_1 z^{-1} + b_2 z^{-2} + b_3 z^{-3})\ u_k \\
&+ (1 + c_1 z^{-1} + c_2 z^{-2} + c_3 z^{-3})\ e_k
\end{aligned}
$$

which can be described by SPM model with the following form:

$$
\left[\begin{array}{ccc|c|c|c}
-z & 0 & 0 & -a_3 & b_3 & c_3 \\
1 & -z & 0 & -a_2 & b_2 & c_2 \\
0 & 1 & -z & -a_1 & b_1 & c_1 \\
\hline
0 & 0 & 1 & -1 & 0 & 1
\end{array}\right]
\begin{bmatrix} x_k \\ y_k \\ u_k \\ e_k \end{bmatrix} = 0
$$

Fuel cell is essentially a nonlinear controlled process, where the parameters and states in the system change at the same time. Therefore, the multiplying of the parameter and the state displays the nonlinear characteristic. The nonlinear problem for joint estimation of the parameters and states can be transformed into a linear estimation issue by using the SPM model.

2) Simultaneous estimation of the states and parameters using the extended Kalman filter in the case of unknown noise information

Suppose $w_k = \left[y_k^T, u_k^T\right]^T$ be known, γ be a column vector containing the elements a_i, b_i in the matrix A^* and B^*, and η be a column vector containing the elements c_i in the matrix C^*. According to equation (4.6), we get

$$\begin{cases} x_{k+1} = E_* x_k + G_* w_k + C_* e_k \\ \quad\;\; = E_* x_k + \tilde{G}_*(w_k) r_k + \tilde{C}_*(e_k)\eta_k \\ 0 = E_0 x_k - J w_k + e_k \end{cases} \tag{4.7}$$

where e_k is the Gaussian white noise with mean 0 and $C_\star$ is the noise parameter matrix, $J = [I_p | 0_{p\times m}]$, $w_k = \left[y_k^T, u_k^T\right]^T$, $\tilde{G}_*(w_k)r = [-A_* | B_*]w_k$. The matrix $\tilde{G}_*(w_k)$ is calculated from the measurement w_k according to the position of the parameters in r. The matrix $\tilde{C}_*(e_k)$ is constructed by the estimated noise in accordance with the specific position of the parameters in η. The key idea of simultaneous estimation of parameters and states is the separation of the unknown parameter a_i, b_i into the vector r and construction of the matrix $\tilde{G}_*(w_k)$ and $\tilde{C}_*(e_k)$ using the measurement w_k.

Let

$$\tilde{F}_k = \begin{bmatrix} E_* & \tilde{G}_*(w_k) & \tilde{C}_*(e_k) \\ 0 & I & 0 \\ 0 & 0 & I \end{bmatrix}, \; s_k = \left[x_k^T \; r_k^T \; \eta_k^T\right]^T, \tilde{H}_k = [E_0 \; 0 \; 0]$$

equation (4.7) can be rewritten as:

$$\begin{cases} s_{k+1} = \tilde{F}_k s_k = f(s_k, w_k) \\ y_k = \tilde{H}_k s_k + e_k = h(s_k) + e_k \end{cases} \tag{4.8}$$

e_k cannot be directly measured but can be estimated from the following equation

$$\hat{e}_k = y_k - \tilde{H}_k \hat{s}_k \tag{4.9}$$

Then, s_k can be estimated by applying an extended Kalman filter, that is to say, the states and parameters will be simultaneously estimated without knowing the parameters in $C_\star$. Introducing the stochastic approximation algorithm [12], the recursive Kalman filter algorithm can be represented by the following equation [13]:

$$\begin{cases} K_k = \Xi_k P_k \tilde{H}_k^T \left(\tilde{H}_k P_k \tilde{H}_k^T + \hat{R}_k\right)^{-1} \\ P_{k+1} = \Xi_k P_k \Xi_k^T - K_k \left(\tilde{H}_k P_k \tilde{H}_k^T + \hat{R}_k\right) K_k^T \\ \tilde{s}_{k+1} = \tilde{F}_k \hat{s}_k + K_k (y_k - \tilde{H}_k \hat{s}_k) \end{cases} \tag{4.10}$$

where,

$$\hat{R}_{k+1} = \hat{R}_k + \frac{1}{k+1}\left(\hat{e}_k\hat{e}_k^T - \hat{R}_k\right) \tag{4.11}$$

$$\Xi_k = \frac{\partial f}{\partial s_k}(s_k, w_k)\Big|_{s_k = \hat{s}_k} \tag{4.12}$$

The partial $\frac{\partial f}{\partial s_k}$ differential can be obtained as follows:

Let

$$T_k = E_* x_k + \tilde{G}_*(w_k)\ r_k + \tilde{C}_*(e_k)\ \eta_k \tag{4.13}$$

From equations (4.7) and (4.8), T_k can be written by

$$T_k = (E_* - C_* E_0)x_k + \tilde{G}_*(w_k)\ r_k + C_* y_k \tag{4.14}$$

where input variable $w_k = \left|y_{k,}^T, u_k^T\right|^T$ is known, then

$$\frac{\partial T_k}{\partial x_k} = E_* - C_* E_0,\ \frac{\partial T_k}{\partial r_k} = \tilde{G}_*(w_k),\ \frac{\partial T_k}{\partial \eta_k} = \tilde{C}_*(e_k) \tag{4.15}$$

and

$$\frac{\partial f}{\partial s_k} = (s_k, w_k) = \begin{bmatrix} \frac{\partial T_k}{\partial s_k} \\ \frac{\partial r_k}{\partial s_k} \\ \frac{\partial \eta_k}{\partial s_k} \end{bmatrix} = \begin{bmatrix} \frac{\partial T_k}{\partial x_k} & \frac{\partial T_k}{\partial r_k} & \frac{\partial T_k}{\partial \eta_k} \\ 0 & I & 0 \\ 0 & 0 & I \end{bmatrix} \tag{4.16}$$

Thus,

$$\frac{\partial f}{\partial s_k}(s_k, w_k) = \begin{bmatrix} E_* - C_* E_0 & \tilde{G}_*(w_k) & \tilde{C}_*(e_k) \\ 0 & I & 0 \\ 0 & 0 & I \end{bmatrix} \tag{4.17}$$

where the matrix Ξ_k is obtained by replacing C_* and e_k with $\tilde{C}_* = C_*|_{\eta = \eta_k}$ and $\hat{e}_k$, respectively.

It can be seen from equation (4.8) that the estimation problem of the parameters and states can be transformed into an extended Kalman filter estimation problem by using the SPM method in the case of unknown noise information.

4.1.2 SPM-based controller design without considering noise estimation information

The structure of the ASFPC algorithm based on SPM without considering the noise estimation information is shown in Figure 4.1:

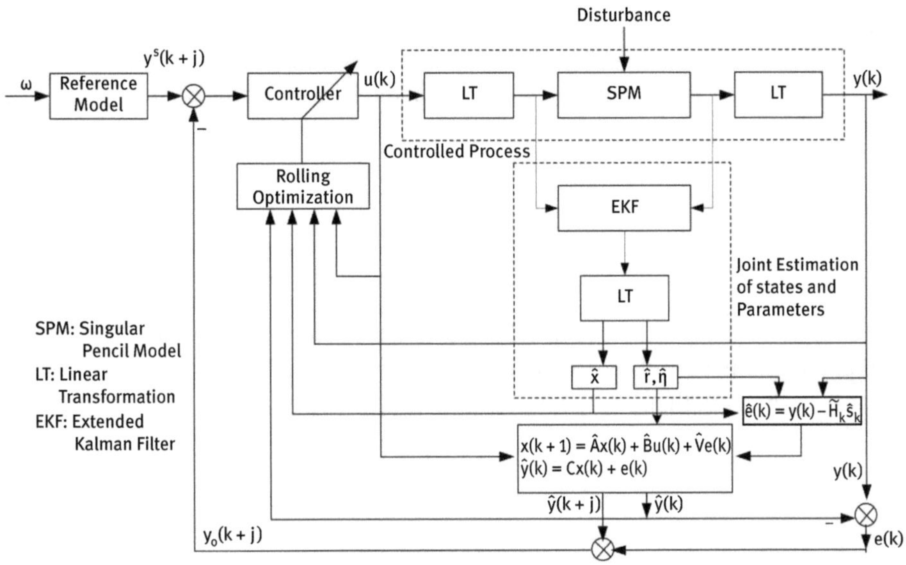

Figure 4.1: Structure of SPM-based ASFPC algorithm without considering noise estimation information.

In order to make full use of the estimated state and parameters, the controlled process is represented by a discrete state space form

$$\begin{cases} x(k+1) = Ax(k) + Bu(k) \\ y(k) = Cx(k) \end{cases} \tag{4.18}$$

where $x \in \mathbb{R}^n$, $y \in \mathbb{R}^r$, $u \in \mathbb{R}^m$, matrix A, B and C is constructed from the estimated parameters in $\hat{s}_k$. From equation (4.18), the predictive states and outputs can be obtained by

$$\hat{x}(k+j) = A^j x(k) + \sum_{i=1}^{j} A^{i-1} Bu(k+j-i) \tag{4.19}$$

$$\hat{y}(k+j) = CA^j x(k) + \sum_{i=1}^{j} CA^{i-1} Bu(k+j-i) \tag{4.20}$$

The predictive outputs can be corrected by feedback.

$$\hat{y}_o(k+j) = \hat{y}(k+j) + y(k) - \hat{y}(k) \tag{4.21}$$

where,

$$\hat{y}(k) = CA^j x(k-j) + \sum_{i=1}^{j} CA^{i-1} Bu(k-i), \quad (j = 1, 2, \ldots, r) \tag{4.22}$$

Different from single-input, single-output systems, the MIMO systems have a prediction horizon p_j for each output, that is, the prediction horizon p is a vector $p = [p_1, p_2, \ldots, p_r]$. From equation (4.20), the predictive value at the sampling time p_j is

$$\hat{y}_j(k+p_j) = C_j A^{p_j} x(k) + \sum_{i=1}^{p_j} C_j A^{i-1} Bu(k+p_j-i) \tag{4.23}$$

$$\hat{y}_j(k) = C_j A^{p_j} x(k-p_j) + \sum_{i=1}^{p_j} C_j A^{i-1} Bu(k-i) \tag{4.24}$$

For a single-valued predictive control algorithm, the control horizon $L = 1$. The control action only changes in the k moment, and remains unchanged in other time, that is, $u(k+i) = u(k), i > 0$, then

$$\hat{y}_j(k+p_j) = C_j A^{p_j} x(k) + S_j(p_j)u(k) \tag{4.25}$$

where $S_j(p_j) = \sum_{i=1}^{P_j} C_j A^{i-1} B$.

The incremental form of the control law is determined by the following objective function, which is the solution of the optimization problems [14, 15], that is, the compromise between the energy of the control input and the error output.

$$J_p = [\hat{y}_o(k+p) - y^s(k+p)]^T Q[\hat{y}_o(k+p) - y^s(k+p)] + \Delta u^T(k)\Lambda\Delta u(k) \tag{4.26}$$

where $y^s(k+p)$ is the reference trajectory and $\hat{y}_o(k+p)$ is the predictive output after the feedback correction.

By taking equations (4.21), (4.24), and (4.25) into equation (4.26), and by letting $\frac{\partial J_p}{\partial k} = 0$, the optimized control action can be obtained.

$$\Delta u(k) = (S^T QS + \lambda)^{-1} S^T Q(y^s(k+p) - y(k) - \Delta_{KX} - \Delta_{SU}) \tag{4.27}$$

where

$$\Delta_{KX} = \begin{bmatrix} C_1 A^{p_1}(x(k) - x(k-p_1)) \\ \vdots \\ C_r A^{p_r}(x(k) - x(k-p_r)) \end{bmatrix}, \quad \Delta_{SU} = \begin{bmatrix} \sum_{i=1}^{p_1}(S_1(p_1) - S_1(i)\Delta u(k-i)) \\ \vdots \\ \sum_{i=1}^{p_r}(S_r(p_r) - S_r(i)\Delta u(k-i)) \end{bmatrix},$$

$$S(p) = \begin{bmatrix} S_1(p_1) \\ \vdots \\ S_r(p_r) \end{bmatrix}, \quad y^s(k+p) = \begin{bmatrix} y^s(k+p_1) \\ \vdots \\ y^s(k+p_r) \end{bmatrix}.$$

In order to adjust the robustness of the controller and limit the magnitude of the control action, a reference trajectory of the following form is adopted.

$$y^s(k+p) = \beta y(k) + (1-\beta)\omega \tag{4.28}$$

where ω is the set point and β is the adjustment factor.

One of the prominent advantages of the state feedback predictive control is the strong ability to suppress unknown disturbance because the state contains not only the control but also the unknown disturbance information. State feedback is equivalent to the feed-forward of the unknown disturbance. Single-valued state feedback predictive control algorithm also has the advantage of small computation, which is particularly suitable for real-time control [16, 17]. ASFPC has a great advantage compared with PID control since the designer can select the cost function according to a specific requirement. The result of the controller design is to minimize the cost function in an automatic way [18–20].

4.1.3 SPM-based controller design considering noise estimation information

The structure of the ASFPC algorithm based on SPM with noise estimation information is shown in Figure 4.2.

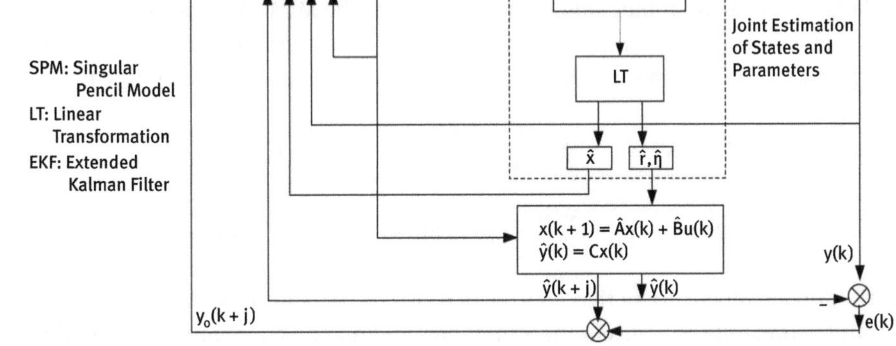

Figure 4.2: Structure of SPM-based ASFPC algorithm considering noise estimation information.

The controlled process is represented by the discrete state space, and the noise information is considered.

$$\begin{cases} x(k+1) = Ax(k) + Bu(k) + Ve(k) \\ y(k) = Cx(k) + e(k) \end{cases} \tag{4.29}$$

where $x \in \mathbb{R}^n, y \in \mathbb{R}^r, u \in \mathbb{R}^m$, matrix A, B and C is constructed from the estimated parameters in $\hat{s}_k$. From equation (4.29), the predictive states and outputs can be obtained by

$$\hat{x}(k+j) = A^j x(k) + \sum_{i=1}^{j} A^{i-1} Bu(k+j-i) + \sum_{i=1}^{j} A^{i-1} V_e(k+j-i) \tag{4.30}$$

$$\begin{aligned} \hat{y}(k+j) = CA^j x(k) + \sum_{i=1}^{j} CA^{i-1} Bu(k+j-i) + e(k+j) \\ + \sum_{i=1}^{j-1} CA^{i-1} Ve(k+j-i) + CA^{j-1} Ve(k) \end{aligned} \tag{4.31}$$

In the right-hand side of equation (4.31), the third and fourth terms are related to the future time, and we can suppose $e(k+j), e(k+j-1), i=1, \cdots, j-1$ are equal to zeros; then, equation (4.31) can be rewritten as

$$\hat{y}(k+j) = CA^j x(k) + \sum_{i=1}^{j} CA^{i-1} Bu(k+j-i) + CA^{j-1} Ve(k) \tag{4.32}$$

The predictive outputs can be corrected by feedback.

$$\hat{y}_o(k+j) = \hat{y}(k+j) + y(k) - \hat{y}(k) \tag{4.33}$$

where,

$$\hat{y}(k) = CA^j x(k-j) + \sum_{i=1}^{j} CA^{i-1} Bu(k-i) + CA^{j-1} Ve(k-j) \tag{4.34}$$

Different from single-input, single-output systems, the MIMO systems have a prediction horizon p_j for each output, that is, the prediction horizon p is a vector $p = [p_1, p_2, \cdots, p_r]$. From equation (4.31), the predictive value at the sampling time p_j is

$$\hat{y}_j(k+p_j) = C_j A^{p_j} x(k) + \sum_{i=1}^{p_j} C_j A^{i-1} Bu(k+p_j-i) + C_j A^{p_j-1} Ve(k) \tag{4.35}$$

$$\hat{y}_j(k) = C_j A^{p_j} x(k-p_j) + \sum_{i=1}^{p_j} C_j A^{i-1} Bu(k-i) + C_j A^{p_j-1} Ve(k-p_j) \tag{4.36}$$

For a single-valued predictive control algorithm, the control horizon is $L=1$. The control action only changes in the k moment, and remains unchanged in other time, that is, $u(k+i) = u(k), i > 0$, then

$$\hat{y}_j(k+p_j) = C_jA^{p_j}x(k) + S_j(p_j)u(k) + C_jA^{p_j-1}Ve(k) \tag{4.37}$$

where $S_j(p_j) = \sum_{i=1}^{p_j} C_jA^{i-1}B$.

The incremental form of the control law is determined by the following objective function, which is the solution of the two optimization problems, that is, the compromise between the energy of the control input and the error output.

$$J_p = [\hat{y}_o(k+p) - y^s(k+p)]^T Q[\hat{y}_o(k+p) - y^s(k+p)] + \Delta u^T(k)\lambda\Delta u(k) \tag{4.38}$$

where $y^s(k+p)$ is the reference trajectory and $\hat{y}_o(k+p)$ is the predictive output after the feedback correction.

By taking equations (4.33), (4.36), and (4.37) into equation (4.38), and by letting $\frac{\partial J_p}{\partial k} = 0$, the optimized control action can be obtained.

$$\Delta u(k) = \left(S^TQS + \lambda\right)^{-1} S^TQ(y^s(k+p) - y(k) - \Delta_{KX} - \Delta_{SU} - \Delta_{TY}) \tag{4.39}$$

where,

$$\Delta_{KX} = \begin{bmatrix} (C_1A^{p_1} - C_1A^{p_1}VC_1)V(x(k) - x(k-p_1)) \\ \vdots \\ (C_rA^{p_r} - C_rA^{p_r}VC_r)V(x(k) - x(k-p_r)) \end{bmatrix}, S_{(P)} = \begin{bmatrix} S_1(p_1) \\ \vdots \\ S_r(p_r) \end{bmatrix},$$

$$\Delta_{SU} = \begin{bmatrix} \sum_{i=1}^{p_1}(S_1(p_1) - S_1(i)\Delta u(k-i)) \\ \vdots \\ \sum_{i=1}^{p_r}(S_r(p_r) - S_r(i)\Delta u(k-i)) \end{bmatrix}, \quad y^s(k+p) = \begin{bmatrix} y_1^s(k+p_1) \\ \vdots \\ y_r^s(k+p_r) \end{bmatrix},$$

$$\Delta_{TY} = \begin{bmatrix} C_1A^{p_j-1}V(y(k) - y(k-p_1)) \\ \vdots \\ C_rA^{p_r-1}V(y(k) - y(k-p_r)) \end{bmatrix}$$

In order to adjust the robustness of the controller and limit the magnitude of the control action, a reference trajectory of the following form is adopted.

$$y^s(k+p) = \beta y(k) + (1-\beta)\omega \tag{4.40}$$

where ω is the set point and β is the adjustment factor.

4.2 Real-time simplified VDFLC

4.2.1 Control structure

The structure of the VDFLC is shown in Figure 4.3, where r is the set point, and K_e, K_{ec}, K_u are the scaling gains of the error e, the error change ec, and the incremental control action Δu, respectively. The physical meaning of e is the error of the set point and the actual output. ec is the error change of the current time and the previous time. Δu is the incremental control action of the fuzzy controller. u is the control action, that is, the input of the actuator. $\alpha(e), \beta(ec), \gamma(e, ec)$ are the flex factors of the domain of e, ec and Δu, which can be utilized to regulate the input and output domain according to the error and error change in order to obtain precise control. The typical two-input (error e and error change ec), one-out (incremental control action Δu) fuzzy controller is adopted. In order to avoid the steady-state error, the integral action is needed. Therefore, an incremental control algorithm is used.

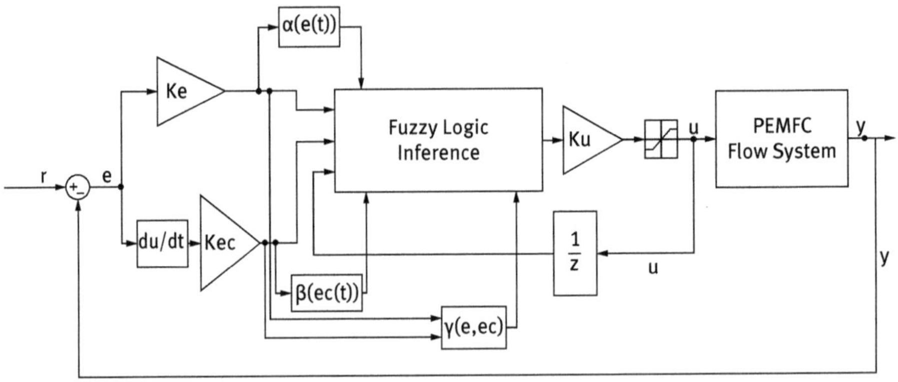

Figure 4.3: Structure of the VDFLC.

4.2.2 Real-time fuzzy inference strategy

The real-time fuzzy inference strategy consists of the following five steps:
- Input normalization
- Fuzzification of inputs
- Rule firing
- Defuzzification of outputs
- Output denormalization

4.2.2.1 Input normalization

The typical fuzzy controllers usually use the error e and the error change ec as the input variables, and control increment Δu as the output variable. The membership functions of inputs and outputs are usually designed off-line in a standardized domain. In our design, the normalized domain for error e is $[-E, +E]$, for error change ec it is $[-EC, +EC]$, and for control increment Δu it is $[-\Delta U, +\Delta U]$. This means that the physical values of the actual controller inputs and outputs can be mapped onto a predefined normalized domain. The mapping of the actual input value onto the input normalized domain is achieved by multiplying the actual physical value and the scaling gain.

X is defined as the normalized domain of e, $X=[-E, +E]$, Y is denoted as the normalized domain of ec, $Y=[-EC, +EC]$, Z is named as the normalized domain of Δu, $Z=[-\Delta U, +\Delta U]$. Then, the variables e and ec can be normalized to x and y.

$$x = K_e \cdot e, \; y = K_{ec} \cdot ec$$

4.2.2.2 Fuzzification

Let $L_e^{(i)}$ be the linguistic variable corresponding to the antecedent normalized variable x in the ith rule, $L_{ec}^{(j)}$ be the linguistic variable corresponding to the antecedent normalized variable y in the jth rule, $L_e^{(i)}$ and $L_{ec}^{(j)}$ are the membership functions of the language variables x and y on the normalized domain X and Y, respectively. $L_e^{(i)}$ and $L_{ec}^{(j)}$ are given by $\mu_{L_e^{(i)}}: X \to [0,1]$ and $\mu_{L_{ec}^{(j)}}: Y \to [0,1]$, respectively.

The normalized inputs x and y adopt triangular membership functions:

$$\mu_{L_e^{(1)}}(x^{(k)}, k) = \begin{cases} 1, & x_1^{(k)} \le x^{(k)} \le x_2^{(k)} \\ 1 - \dfrac{x^{(k)} - x_2^{(k)}}{x_3^{(k)} - x_2^{(k)}}, & x_2^{(k)} \le x^{(k)} \le x_3^{(k)} \\ 0, & \text{else} \end{cases} \tag{4.41}$$

$$\mu_{L_e^{(i)}}(x^{(k)}, k) = \begin{cases} 1 + \dfrac{x^{(k)} - x_i^{(k)}}{x_i^{(k)} - x_{i-1}^{(k)}}, & x_{i-1}^{(k)} \le x^{(k)} \le x_i^{(k)} \\ 1 - \dfrac{x^{(k)} - x_i^{(k)}}{x_{i+1}^{(k)} - x_i^{(k)}}, & x_i^{(k)} \le x^{(k)} \le x_{i+1}^{(k)} \\ 0, & \text{else} \end{cases} \tag{4.42}$$

$$\mu_{L_e^{(p)}}(x^{(k)}, k) = \begin{cases} 0, & \text{else} \\ 1 + \dfrac{x^{(k)} - x_{p+1}^{(k)}}{x_{p+1}^{(k)} - x_p^{(k)}}, & x_p^{(k)} \le x^{(k)} \le x_{p+1}^{(k)} \\ 1, & x_{p+1}^{(k)} \le x^{(k)} \le x_{p+2}^{(k)} \end{cases} \tag{4.43}$$

$$\mu_{L_{ec}^{(1)}}(y^{(k)},k)=\begin{cases}1, & y_1^{(k)}\le y^{(k)}\le y_2^{(k)}\\ 1-\dfrac{y^{(k)}-y_2^{(k)}}{y_3^{(k)}-y_2^{(k)}}, & y_2^{(k)}\le y^{(k)}\le y_3^{(k)}\\ 0, & \text{else}\end{cases} \tag{4.44}$$

$$\mu_{L_{ec}^{(j)}}(y^{(k)},k)=\begin{cases}1+\dfrac{y^{(k)}-y_j^{(k)}}{y_j^{(k)}-y_{j-1}^{(k)}}, & y_{j-1}^{(k)}\le y^{(k)}\le y_j^{(k)}\\ 1-\dfrac{y^{(k)}-y_j^{(k)}}{y_{j+1}^{(k)}-y_j^{(k)}}, & y_j^{(k)}\le y^{(k)}\le y_{j+1}^{(k)}\\ 0, & \text{else}\end{cases} \tag{4.45}$$

$$\mu_{L_{ec}^{(q)}}(y^{(k)},k)=\begin{cases}0, & \text{else}\\ 1+\dfrac{y^{(k)}-y_{q+1}^{(k)}}{y_{q+1}^{(k)}-y_q^{(k)}}, & y_q^{(k)}\le y^{(k)}\le y_{q+1}^{(k)}\\ 1, & y_{q+1}^{(k)}\le y^{(k)}\le y_{q+2}^{(k)}\end{cases} \tag{4.46}$$

where $-E=x_1<x_2<\cdots<x_{p+2}=+E$, $-EC=y_1<y_2<\cdots<y_{q+2}=+EC$.

4.2.2.3 Fuzzy reasoning

It can be seen from the triangular membership function that any input belongs to a maximum of two membership functions. Therefore, for a two-input system, at most four control rules can be activated for each cycle [21, 22]. Specifically, suppose (x^k,y^k) is the normalized input variable, the membership function belonging to x is $\mu_{L_e^{(i)}}(x^k,k)$ and $\mu_{L_e^{(i+1)}}(x^k,k)$, the membership function belonging to y is $\mu_{L_{ec}^{(j)}}(y^k,k)$ and, $\mu_{L_{ec}^{(j+1)}}(y^k,k)$, $\mu_{L_{ec}^{(i)}}$, $\mu_{L_{ec}^{(i+1)}}$, $\mu_{L_{ec}^{(j)}}$, and $\mu_{L_{ec}^{(j+1)}}$ is the corresponding membership. Then, four control rules $L_{\Delta u}^{(i,j)}$, $L_{\Delta u}^{(i,j+1)}$, $L_{\Delta u}^{(i+1,j)}$, $L_{\Delta u}^{(i+1,j+1)}$ will be fired. Thus, the corresponding four "if – then" rules can be obtained:

If x is $L_e^{(i)}$, **and** y is $L_{ec}^{(j)}$, **then** Δu is $L_{\Delta u}^{(i,j)}$.
If x is $L_e^{(i)}$, **and** y is $L_{ec}^{(j+1)}$, **then** Δu is $L_{\Delta u}^{(i,j+1)}$.
If x is $L_e^{(i+1)}$, **and** y is $L_{ec}^{(j)}$, **then** Δu is $L_{\Delta u}^{(i+1,j)}$.
If x is $L_e^{(i+1)}$, **and** y is $L_{ec}^{(j+1)}$, **then** Δu is $L_{\Delta u}^{(i+1,j+1)}$.

The normalized input variables x and y of the fuzzy controller are associated with a series of linguistic variables, and different linguistic variable pair (x, y) decisions control action Δu. Suppose that the output domain is divided into $c_m^{(k)}$, $m=1,2,\cdots,n$, satisfying $-\Delta U=c_1<c_2<\cdots<c_n=+\Delta U$, and a series of linguistic variables are connected to the output Δu. Thus, the membership functions of output variable Δu can be expressed as a two-dimensional table between $c_m^{(k)}$ and the linguistic variables.

Look up the membership function table of variable Δu, four $1 \times n$ membership function matrixes $L_{\Delta u}^{(i,j)}, L_{\Delta u}^{(i,j+1)}, L_{\Delta u}^{(i+1,j)}$, and $L_{\Delta u}^{(i+1,j+1)}$ can be obtained.

Let ^ be a minimum operator, and compute

$$\tilde{\mu}^{(m)}_{L_{\Delta u}^{(i,j)}} = \wedge\left(\mu_{L_e^{(i)}}, \mu_{L_{ec}^{(j)}}, \mu^{(m)}_{L_{\Delta u}^{(i,j)}}\right),$$

$$\tilde{\mu}^{(m)}_{L_{\Delta u}^{(i,j+1)}} = \wedge\left(\mu_{L_e^{(i)}}, \mu_{L_{ec}^{(j+1)}}, \mu^{(m)}_{L_{\Delta u}^{(i,j+1)}}\right),$$

$$\tilde{\mu}^{(m)}_{L_{\Delta u}^{(i+1,j)}} = \wedge\left(\mu_{L_e^{(i+1)}}, \mu_{L_{ec}^{(j)}}, \mu^{(m)}_{L_{\Delta u}^{(i+1,j)}}\right),$$

$$\tilde{\mu}^{(m)}_{L_{\Delta u}^{(i+1,j+1)}} = \wedge\left(\mu_{L_e^{(i+1)}}, \mu_{L_{ec}^{(j+1)}}, \mu^{(m)}_{L_{\Delta u}^{(i+1,j+1)}}\right),$$

where m = 1,2,. . ., n.

Let $\vee$ be a maximum operator, and calculate

$$\tilde{\tilde{\mu}}^{(m)}_{\Delta u} = \vee\left(\tilde{\mu}^{(m)}_{L_{\Delta u}^{(i,j)}}, \tilde{\mu}^{(m)}_{L_{\Delta u}^{(i,j+1)}}, \tilde{\mu}^{(m)}_{L_{\Delta u}^{(i+1,j)}}, \tilde{\mu}^{(m)}_{L_{\Delta u}^{(i+1,j+1)}}\right)$$

where m = 1,2,. . ., n.

4.2.2.4 Defuzzification

The normalized incremental control action Δu is determined using the defuzzification method of the center of gravity in equation (4.47).

$$\Delta u = \frac{\sum_{m=1}^{n} \tilde{\tilde{\mu}}^{(m)}_{\Delta u} * c_m}{\sum_{m=1}^{n} \tilde{\tilde{\mu}}^{(m)}_{\Delta u}} \tag{4.47}$$

4.2.2.5 Output denormalization

The denormalization is achieved by multiplying the normalized output value by the scaling gain. Then the control action $u^{(k)}$ can be calculated by:

$$u^{(k)} = u^{(k-1)} + K_u \cdot \Delta u^{(k)} \tag{4.48}$$

Repeat the above steps until the program is finished.

4.2.2.6 Variable domain strategy

VDFLC can achieve precise control by adjusting the input and output domains with less control rules [23]. The core idea is that the domain shrinks with the decrease of the error, based on the unchanged form of control rules. The contraction of the domain is equivalent to the increase of the control rules. Thus, the control precision can be improved. In this case, in the expert's experience, the division of the region and the selection of membership functions become less important, as long as mastering the general trend of the control rules [24–32]. Figure 4.4 shows the contraction and expansion of the domain. The change of domain is achieved by multiplication with the scaling factor. It is defined as follows:

$$\alpha(x) = \left[\frac{|x|}{E}\right]^{\tau_1}, 0 < \tau_1 < 1 \tag{4.49}$$

$$\beta(y) = \left[\frac{|y|}{EC}\right]^{\tau_2}, 0 < \tau_2 < 1 \tag{4.50}$$

$$\gamma(x, y) = \left[\left[\frac{|x|}{E}\right]^{\tau_1} \cdot \left[\frac{|y|}{EC}\right]^{\tau_2}\right]^{\tau_3}, 0 < \tau_1, \tau_2, \tau_3 < 1 \tag{4.51}$$

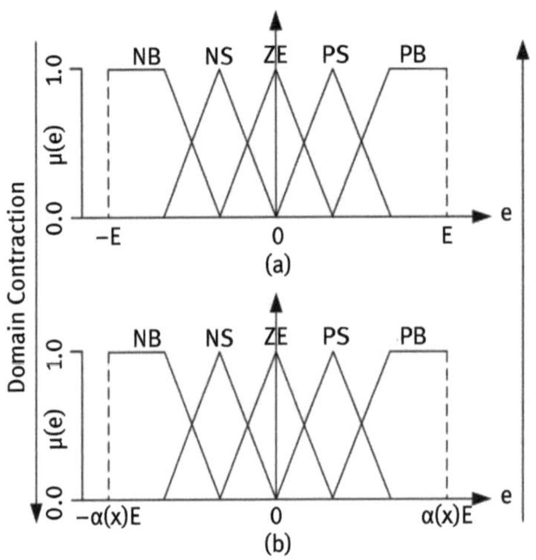

Figure 4.4: Domain contraction and expansion.

Therefore, the algorithm of variable domain can be realized by the following steps [2,27]:

Step 0: Initialize the normalized domain of e, ec, and Δu, $X = [-E, +E], Y = [-EC, +EC]$, $Z = [-\Delta U, +\Delta U]$. According to the initial input $x^{(0)} \in X$, $y^{(0)} \in Y$ and the real-time fuzzy logic reasoning method to calculate $u^{(1)}$.

Step 1: The error $e^{(1)}$ can be obtained by applying $u^{(1)}$ to the controlled process, comparing the process output to the reference; then the error change $ec^{(1)}$ can also be obtained by difference. Normalized inputs $x^{(1)}$ and $y^{(1)}$ can be obtained by input normalizations. Next, computing $x_i^{(1)} = \alpha(x^{(1)})\, x_i^{(0)}$, $y_j^{(1)} = \beta\,(y^{(1)})\, y_j^{(0)}$,

$c_m^{(1)} = \gamma(x^{(1)}, y^{(1)})c_m^{(0)}$, $u^{(2)}$ can be calculated by means of fuzzification, fuzzy reasoning, defuzzification, and denormalization of outputs.

...

Step k: The error $e^{(k)}$ can be obtained by applying $u^{(k)}$ to the controlled process, comparing the process output to the reference; then the error change $ec^{(k)}$ can also be obtained by difference. Normalized inputs $x^{(k)}$ and $y^{(k)}$ can be obtained by input normalizations. Next, computing $x_i^{(k)} = \alpha(x^{(k)})\, x_i^{(0)}$, $y_j^{(k)} = \beta(y^{(k)})\, y_j^{(0)}$, $c_m^{(k)} = \gamma(x^{(k)}, y^{(k)})c_m^{(0)}$, $u^{(k+1)}$ can be derived from the real-time fuzzy inference approach.

4.3 Simulations and experiments

4.3.1 SPM-based ASFPC

4.3.1.1 Simulations

In order to evaluate the SPM-based ASFPC method, we have carried out simulative and experimental researches. The proton exchange membrane (PEM) fuel cell flow system for simulation has been established in the MATLAB/Simulink environment (see Figure 4.5). The SPM-based adaptive state feedback predictive controller is utilized to control the cathode pressure while the PI controller is implemented to control the anode pressure. The cathode pressure changes with the set point, and the anode pressure follows the changes of the cathode pressure. To avoid the generation of algebraic rings, the MEMORY module is used. The parameter estimation algorithm and the ASFPC algorithm are written by Simulink's C-S function. The anode and cathode flow models can be identified by the SPM-based state and parameter estimation method. In the parameters estimation, the given voltage of the air flowmeter is the input variable and the cathode pressure is the output variable. The range of the given voltage of the air flow meter is 0 to 5V. To sufficient excitation of the system, the random number between 0 and 5V changes is treated as input. The random input together with the output data of the corresponding cathode pressure is used to identify the cathode flow model. Similarly, the anode flow model will also be obtained according to the given voltages of the hydrogen flowmeter and the corresponding output data of the anode pressure.

$$\begin{cases} x_{k+1} = \begin{bmatrix} 0 & -0.6338 \\ 1 & 1.5956 \end{bmatrix} x_k + \begin{bmatrix} -0.9472 \\ 1.4163 \end{bmatrix} u_k + \begin{bmatrix} -0.1787 \\ -0.0201 \end{bmatrix} e_k \\ y_k = \begin{bmatrix} 0 & 1 \end{bmatrix} x_k + e_k \qquad \text{Cathode Model} \end{cases}$$

$$\begin{cases} x_{k+1} = \begin{bmatrix} 0 & -0.3855 \\ 1 & 1.3509 \end{bmatrix} x_k + \begin{bmatrix} -0.6514 \\ 0.8933 \end{bmatrix} u_k + \begin{bmatrix} -0.4935 \\ -0.0466 \end{bmatrix} e_k \\ y_k = \begin{bmatrix} 0 & 1 \end{bmatrix} x_k + e_k \qquad \text{Anode Model} \end{cases}$$

The control performance of the simulation is shown in Figure 4.6. Compared with the subsequent experimental results, the simulation basically reflects the dynamic of the process.

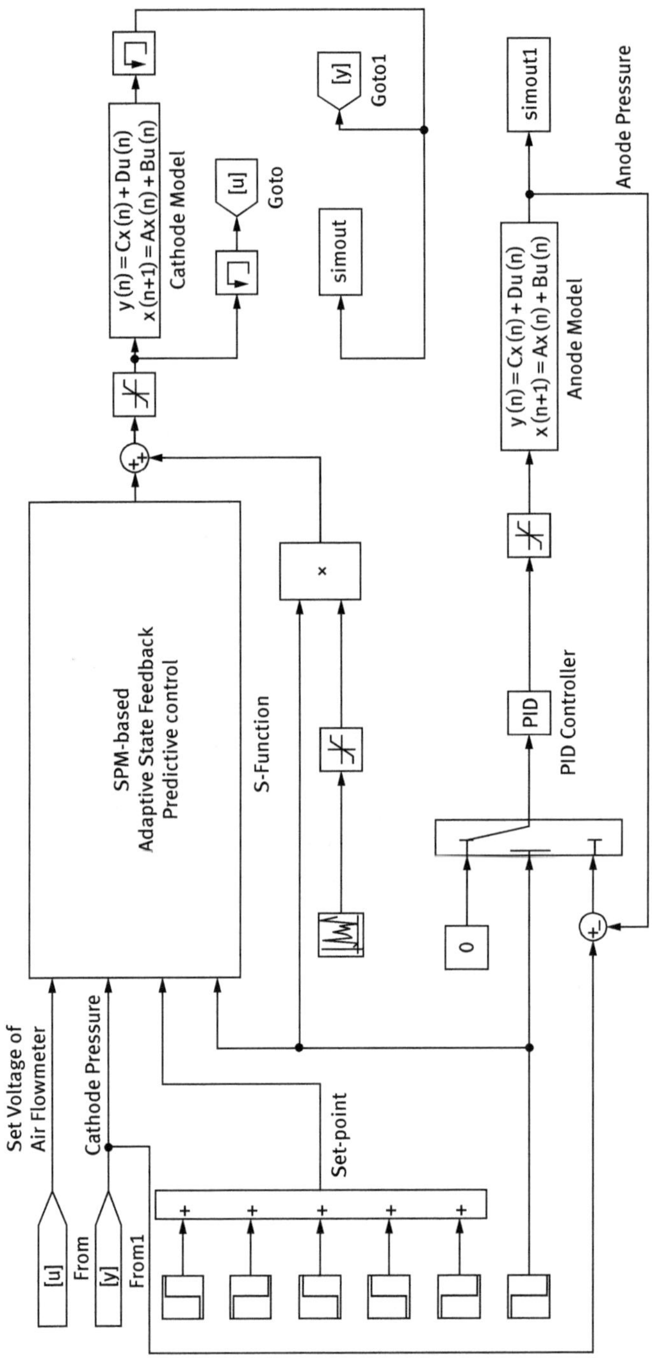

Figure 4.5: Control strategy simulative realization of PEM fuel cell flow system in Simulink.

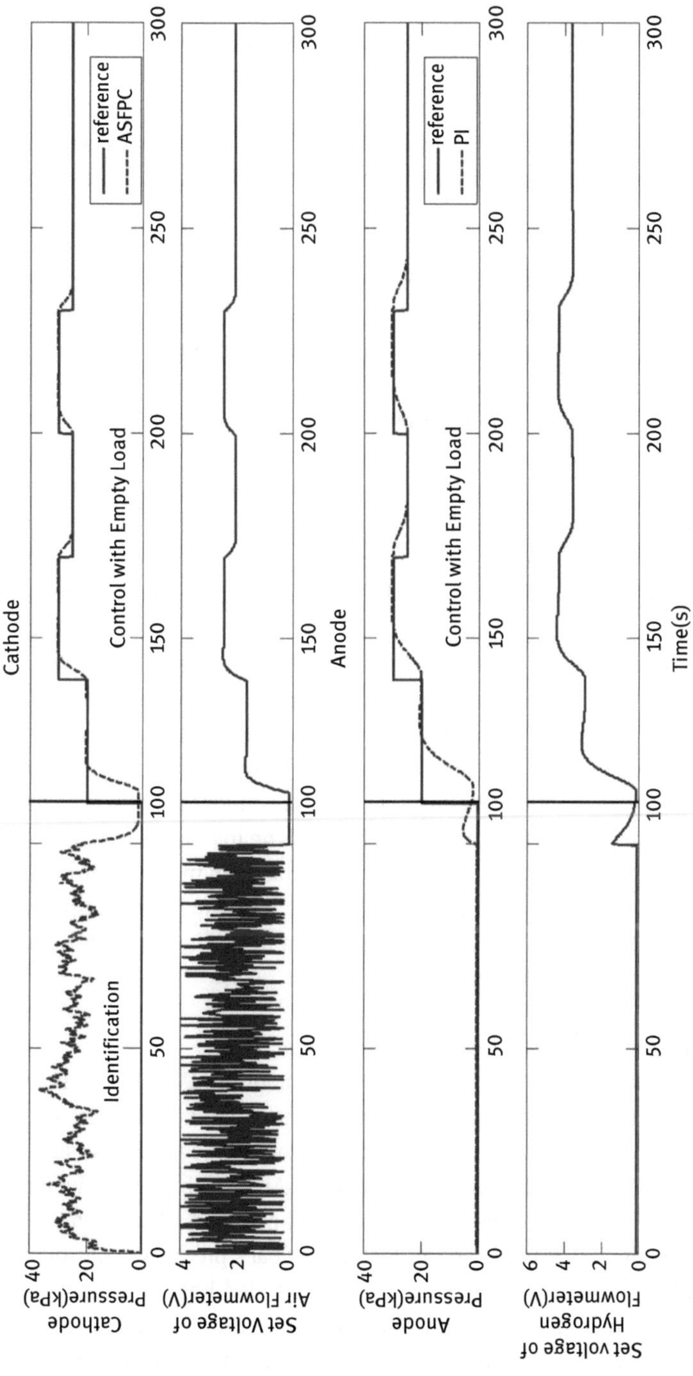

Figure 4.6: Simulation control performance of the PEM fuel cell flow system.

4.3.1.2 Real-time applications

The cathode flow system for the PEM fuel cell is a single-input, single-output system. The input is the setting voltage of the air flowmeter, and the output is the cathode pressure. The anode flow system is also a single-input, single-output system. The input is the setting voltage of the hydrogen flowmeter. The output is the anode pressure. The goal of the control for the flow system is to automatically and quickly adjust the gas flow of the anode and cathode to maintain the stack pressure in a certain range, while ensuring that hydrogen and air flow to follow the load changes.

1) Control strategy

In order to clarify the control strategy, it is necessary to elaborate introduce the related hardware. In our test rig, the inlet of the cathode passage is equipped with an air flow meter, and the inlet of the anode passage is equipped with a hydrogen flow meter. The input voltage of the air flow meter is $0 \sim 5$V, and the output flow range is 0 ~ 100 SLM. The input voltage of the hydrogen flow meter is also $0 \sim 5$V, while the output flow range is 0 ~ 20 SLM.

Two hand valves, which can be adjusted but maintained at the same degree of opening in each experiment, are equipped in the outlet pathway of the anode and the cathode, respectively. The pressure sensors are mounted on the outlets of the anode and the cathode. In order to avoid damage to the PEM, the control goal is to maintain the anode and cathode pressure balance. Due to the nonlinear chemical reaction between the anode and cathode, uncertainties of the disturbance and time lag, PID controller cannot cope well. The cathode pressure adopts the SPM-based ASFPC, by adjusting the air flow to control the cathode pressure at a given point, and anode control adopts the conventional PI controller, by adjusting the hydrogen flow to control the anode pressure to follow the cathode pressure changes. Thus, the change in load can be viewed as a disturbance to the system. If the load increases, the stack will consume more air and hydrogen. In order to maintain the pressure, it is necessary to increase the given voltage of the air and hydrogen flowmeter to supply more air and hydrogen. The goal of the ASFPC is to maintain a more stable pressure, while meeting the power requirements of the load.

2) Achieve control strategy in the Simulink environment

The control module is built up in the Simulink environment of MATLAB as shown in Figure 4.7. The ASFPC method based on SPM is used to control the cathode pressure, and the PI controller with $P = 0.06$, and $I = 0.04$ is utilized to control the anode pressure. The 'ADCSFun' and 'DACSFun' modules are written in MATLAB's C-S functions, which are the interfaces between controller, external sensors, and actuators. The 'SaveCC' is a storage module. S-function 'SPM-based adaptive state feedback predictive control' is the core algorithm written in C-S function, including the extended Kalman filter for parameter estimation and the adaptive state feedback predictive controller. In order to switch the identification and control algorithm, a step module is used by reading the state of the

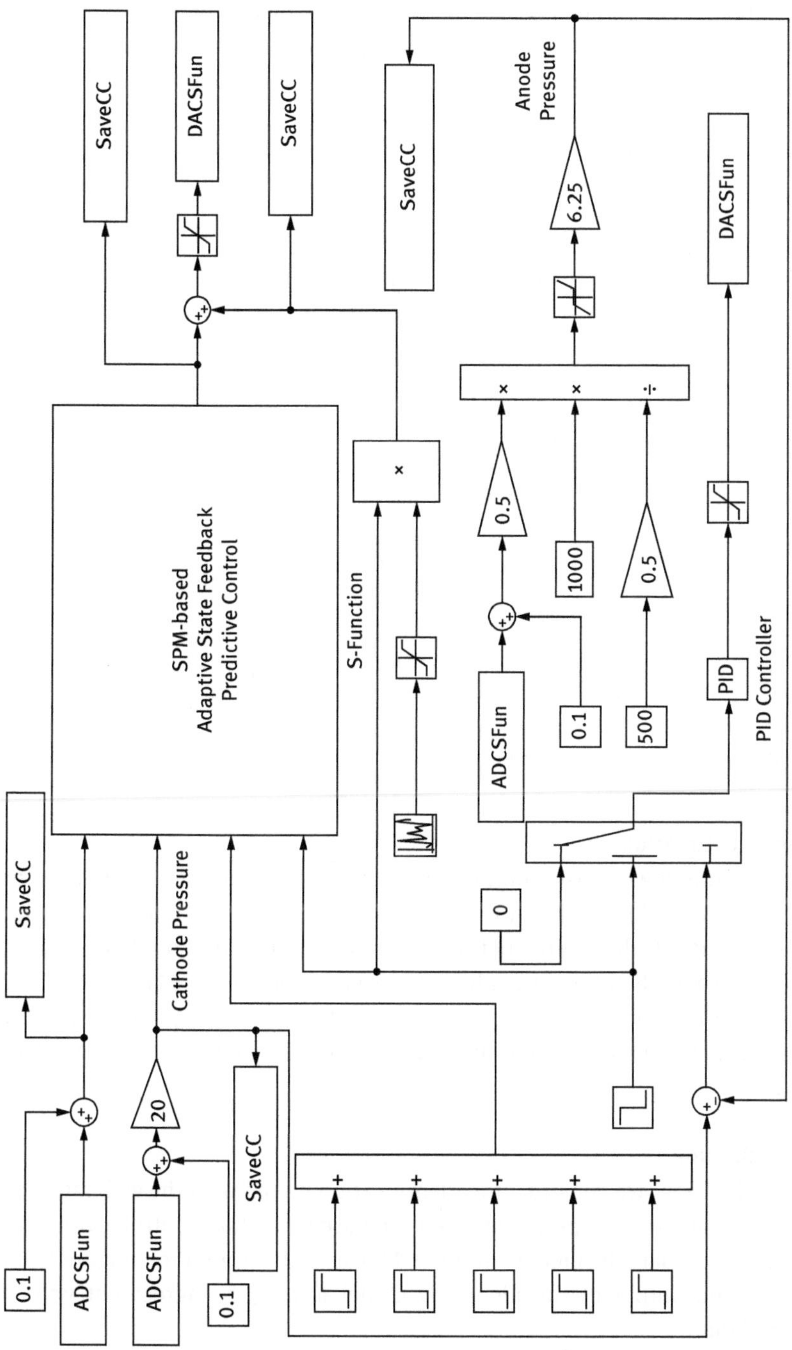

Figure 4.7: Implementation of PEM fuel cell flow system control strategy in Simulink.

step module in the identification and control algorithm written in the C-S function. When it is equal to 1, the identification algorithm is executed first; when it changes to 0, the control algorithm is then executed. The prediction horizon is 4, the sampling time is 0.2 seconds, $\lambda = 600$, $\beta = 0$. Figure 4.8 is the result of the estimation of the parameters and states. Figure 4.7 can be converted into executable code to run in the controller's ARM (Advanced RISC (Reduced Instruction Set Computing) Machines) processor.

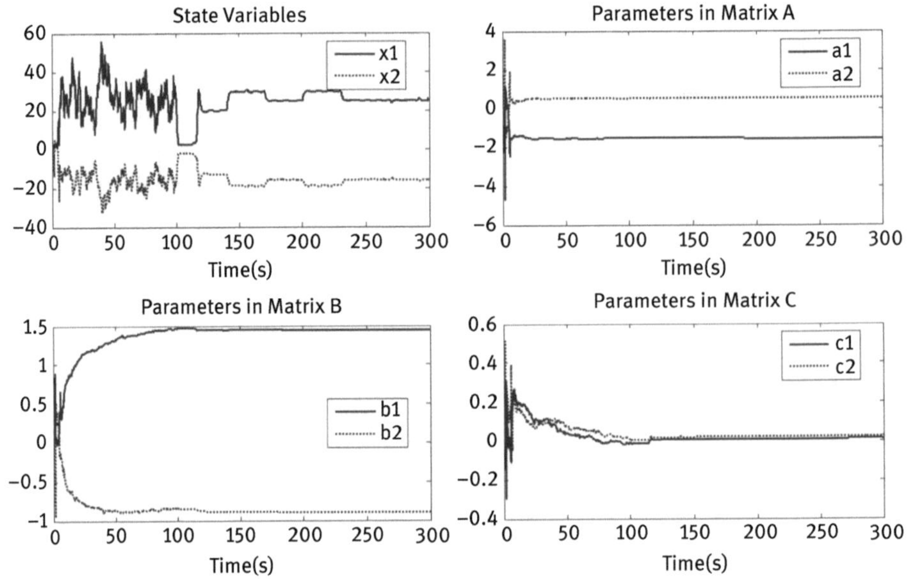

Figure 4.8: Parameters and states estimation.

3) Parameter estimation

In the case of unknown noise information, simultaneous estimation of states and parameters can be achieved using an extended Kalman filter. In order to obtain a more accurate model, the online identification algorithm is first implemented, and then the adaptive predictive control algorithm is executed after the 110th second and the online identification algorithm continues to run until the program terminates. It can be seen from Figure 4.8 that the states and parameters in the iteration are always changing.

4) Model validation

For model validation, we construct the identification model with the estimated state and parameters at each sampling time, and then use the actual control action as the model input to compare the model output and the actual output. From Figure 4.9, we can see that the output of the identified model is in good agreement with the experimental data.

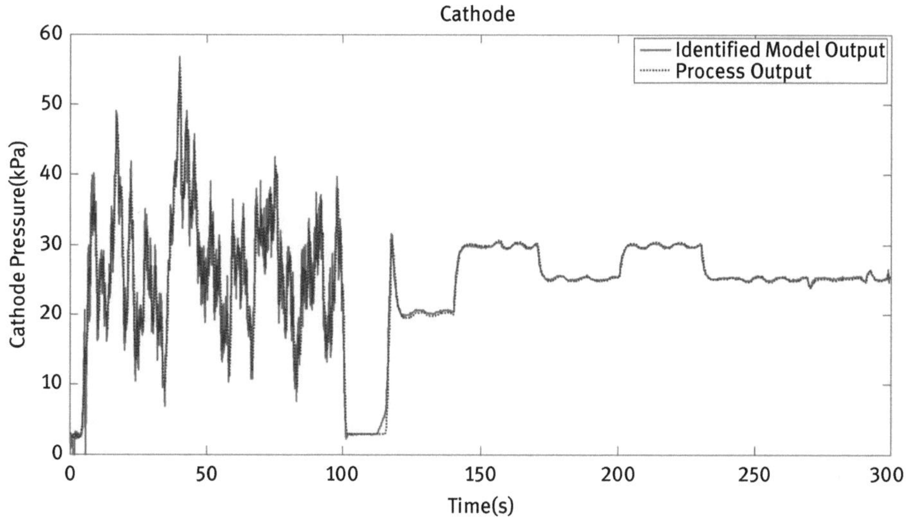

Figure 4.9: Identified model output and actual process output comparison.

5) Control with empty load and a certain load

Figures 4.10 and 4.11 show the control performance and control actions for PEM fuel cell flow system with an empty load (between the 110th second and the 250th second) and under certain loads (after the 250th second). The dynamic process of simulation and experiment is roughly the same compared with the simulation results in Figure 4.6. The load varies from 0 W to 445W and is almost equal to half the maximum

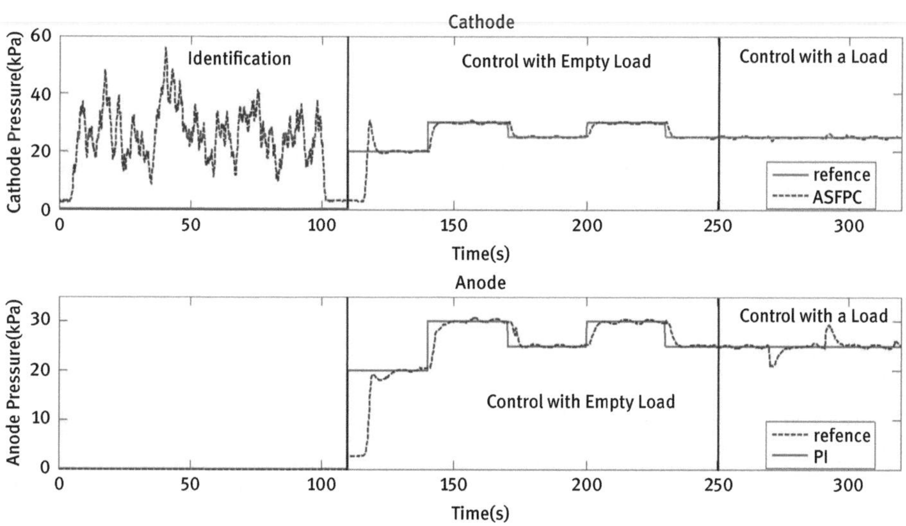

Figure 4.10: PEM fuel cell flow system online control.

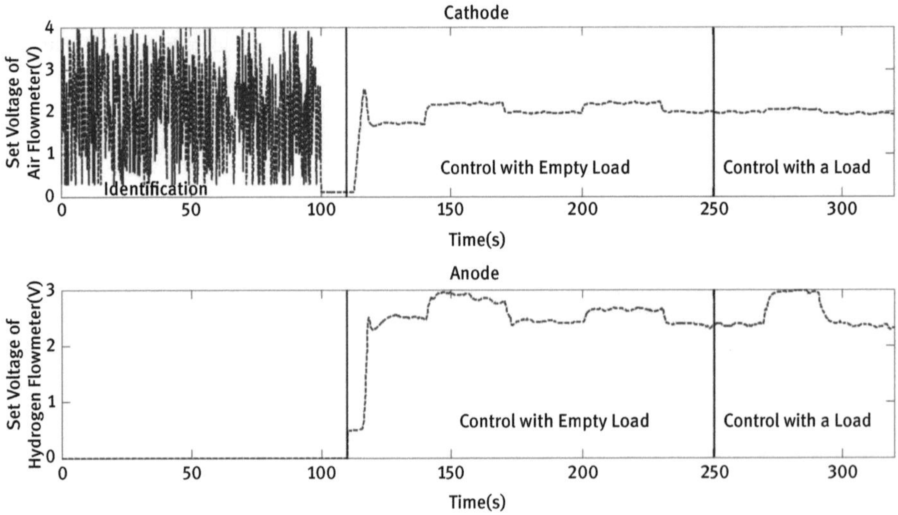

Figure 4.11: PEM fuel cell flow system anode and cathode controller output.

output of the stack. The change in the load is equivalent to the perturbation of the pressure control of the PEM fuel cell flow system. To compare the disturbance suppression of the SPM-based ASFPC (Exp.1), we have done another two experiments: one is the predictive control without state feedback (Exp.2) and the other is the PI control (Exp.3). From Figures 4.12 and 4.13, we can see that the SPM-based ASFPC algorithm

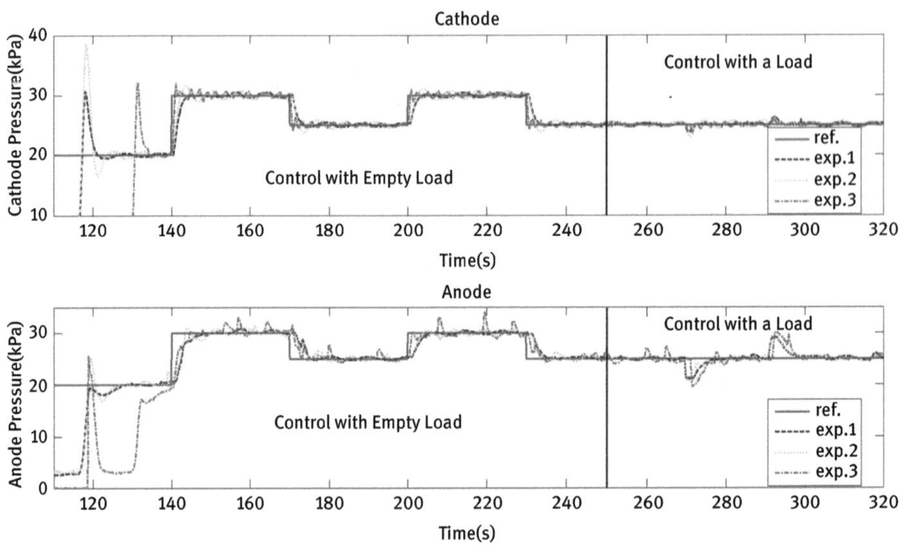

Figure 4.12: Performance comparison of three control methods.

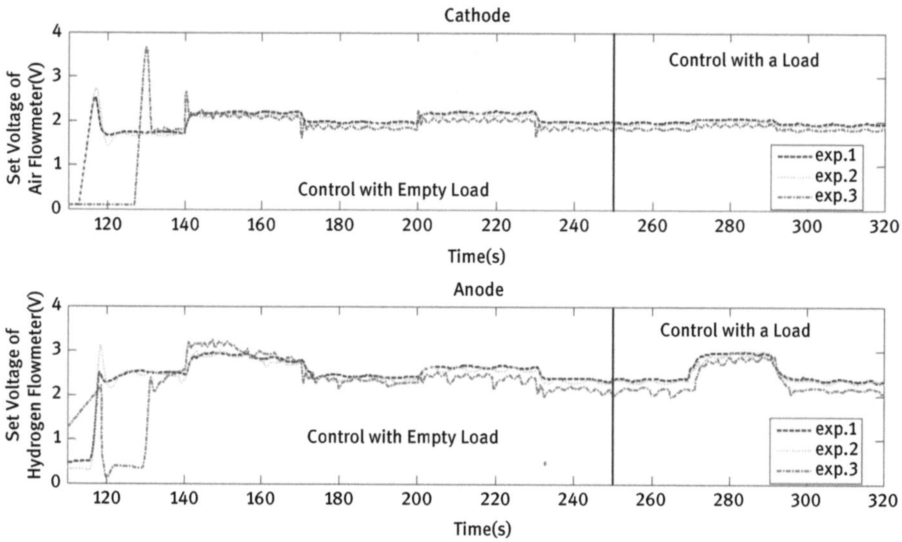

Figure 4.13: Control actions comparison of the three control methods.

has better ability to suppress the disturbance than the other two algorithms, because the SPM-based ASFPC algorithm takes into account both state and parameter information. Except the disturbance of the load, the block to the air passage incurred by the water generated in reaction is also a disturbance to the cathode pressure.

4.3.2 Real-time simplified variable-domain fuzzy control

1) Fuzzy controller design

The initial normalized domain of input e and ec is $[-2, +2]$, and the initial normalized domain of the output Δu is $[-3, +3]$. Note that the initial normalized domain of e and ec can be different. The choice of the initial normalized domain is somewhat arbitrary because the input range can be adjusted by the scaling gains, that is, the actual signals of inputs e and ec can be mapped to the prenormalized domain $[-2, +2]$ by multiplying them with the scaling gains K_e and K_{ec}. The normalized output can be mapped to the actual physical domain by denormalizing process. Thus, the actual output value is equal to the output value of the previous step plus the product of the normalized output value and the scaling gain K_u (see equation (4.48)).

The input variable e corresponds to a set of language control variables

$$L(e) = \{NB, NS, ZE, PS, PB\}$$

where, NB represents Negative Big, NS represents Negative Small, ZE represents Zero, PS represent Positive Small, and PB represents Positive Big.

The membership function of the variable e on the normalized domain $[-2, +2]$ is shown in Figure 4.14. It can be seen that each input at most belongs to two sets, and each membership degree is a linear function of the input.

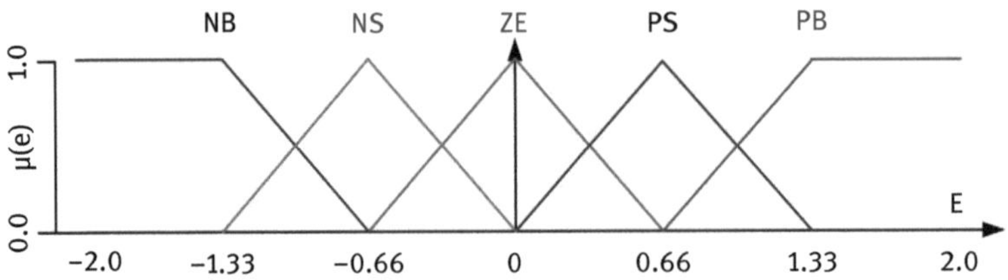

Figure 4.14: The membership function of the variable e on the normalized domain $[-2, +2]$.

The membership function of the input variable ec on the normalized domain $[-2, +2]$ can be constructed in the same way, which is shown in Figure 4.15.

$$L(ec) = \{NB, NM, NS, ZE, PS, PM, PB\}$$

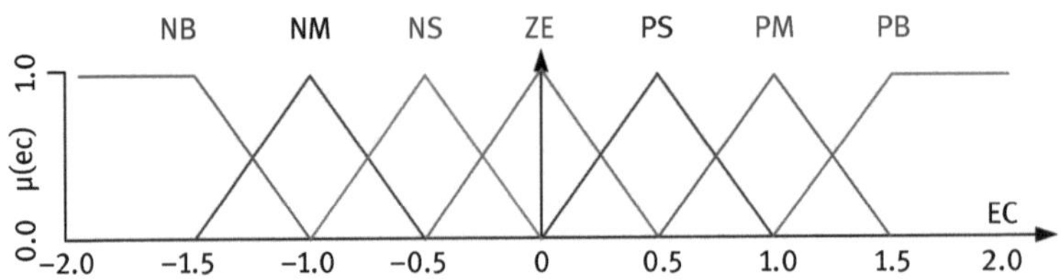

Figure 4.15: The membership function of the variable ec on the normalized domain $[-2, +2]$.

where NM and PM denote Negative Medium and Positive Medium, respectively.

In the variable domain strategy, the domain can be contracted as the error decreases. In this mechanism, the designer can use less membership function partitions of the error e and the error change ec to obtain the same or even better control performance than the ordinary fuzzy controller with more membership function partitions.

To simplify the defuzzification process, the output membership function of Δu uses the discrete form shown in Table 4.1.

The typical dynamic step response of the control system can be shown as in Figure 4.16. It can be seen that the y-axis can indicate the error between the reference and actual output and the slope of the curve can represent the error change. Therefore, the control rules can be designed according to this dynamic response curve.

Table 4.1: The membership function of the output Δu on normalized domain [−3, +3].

	−3.0	−2.0	−1.0	0	1	2	3
NB	1	0.7	0.3	0	0	0	0
NM	0.7	1	0.7	0.3	0	0	0
NS	0.3	0.7	1	0.7	0.3	0	0
ZE	0	0.3	0.7	1	0.7	0.3	0
PS	0	0	0.3	0.7	1	0.7	0.3
PM	0	0	0	0.3	0.7	1	0.7
PB	0	0	0	0	0.3	0.7	1

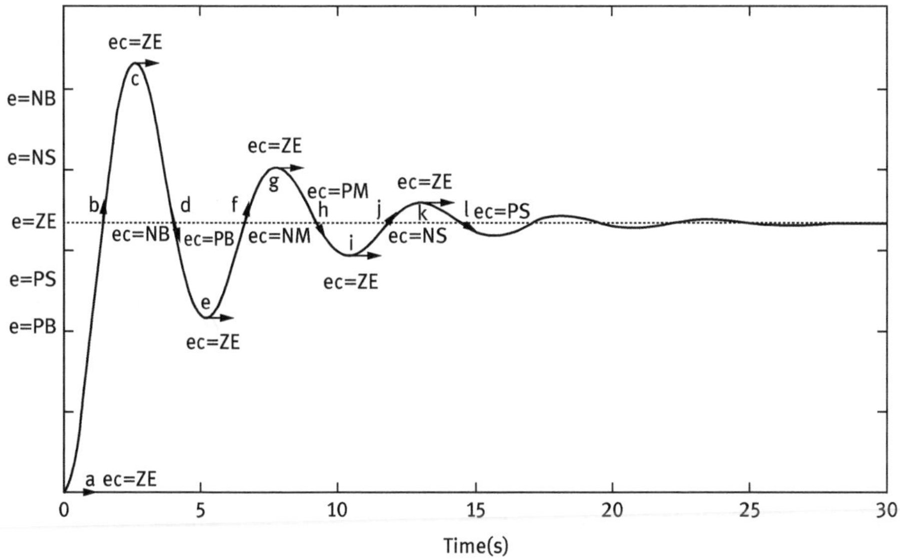

Figure 4.16: Dynamic response of the controlled system under step input.

In fact, the only consideration of the points a, b, ..., l is sufficient to control the process. Table 4.2 lists the corresponding relationships between reference points and control rules. The entire control rules are shown in Table 4.3. Its regulation process can be explained by the fact that, at the beginning of the control procedure, the input e can fall into the ZE region by adjusting the scaling gain K_e, and then the variable domain strategy begins to work, the error e can remain near the ZE region, where the adjustment of the variable ec is very efficient. Sometimes, e may fall into the neighborhood of ZE, however, since ZE and adjacent membership function regions of NS, PS are overlaping, and the control action is not zero. In the role of negative feedback, e will be pulled back to the ZE area, through several control cycles, and the control objectives can be achieved.

Table 4.2: Relationship between reference points and control rules.

Rule No.	e	ec	Δu	Reference Points
1	PB	ZE	PB	a
2	PS	ZE	PS	e/i
3	ZE	PB	PB	d
4	ZE	PM	PM	h
5	ZE	PS	PS	l
6	NB	ZE	NB	c
7	NS	ZE	NS	g/k
8	ZE	NB	NB	b
9	ZE	NM	NM	f
10	ZE	NS	NS	j
11	ZE	ZE	ZE	Set point

Table 4.3: Control rules.

	Δu	e				
		NB	**NS**	**ZE**	**PS**	**PB**
	NB	ZE	ZE	NB	ZE	ZE
	NM	ZE	ZE	NM	ZE	ZE
	NS	ZE	ZE	NS	ZE	ZE
e	**ZE**	NB	NS	ZE	PS	PB
	PS	ZE	ZE	PS	ZE	ZE
	PM	ZE	ZE	PM	ZE	ZE
	PB	ZE	ZE	PB	ZE	ZE

2) Simulations

In this section, we use the model identified in Section 4.3.1 as the controlled process to study some of the performances of this algorithm. The simulation diagram of the entire flow control system is constructed in MATLAB/Simulink as shown in Figure 4.17, and the simulation control performance is shown in Figure 4.18.

3) Real-time applications

i. Control strategy achievement in the Simulink environment

Constructing the control strategy in MATLAB/Simulink environment as shown in Figure 4.18, the cathode pressure varies with the set point and the anode pressure follows the change of the cathode pressure. The control of the cathode pressure adopts the real-time variable domain fuzzy controller and the control of the anode pressure implements the real-time fixed domain fuzzy controller. The sampling time

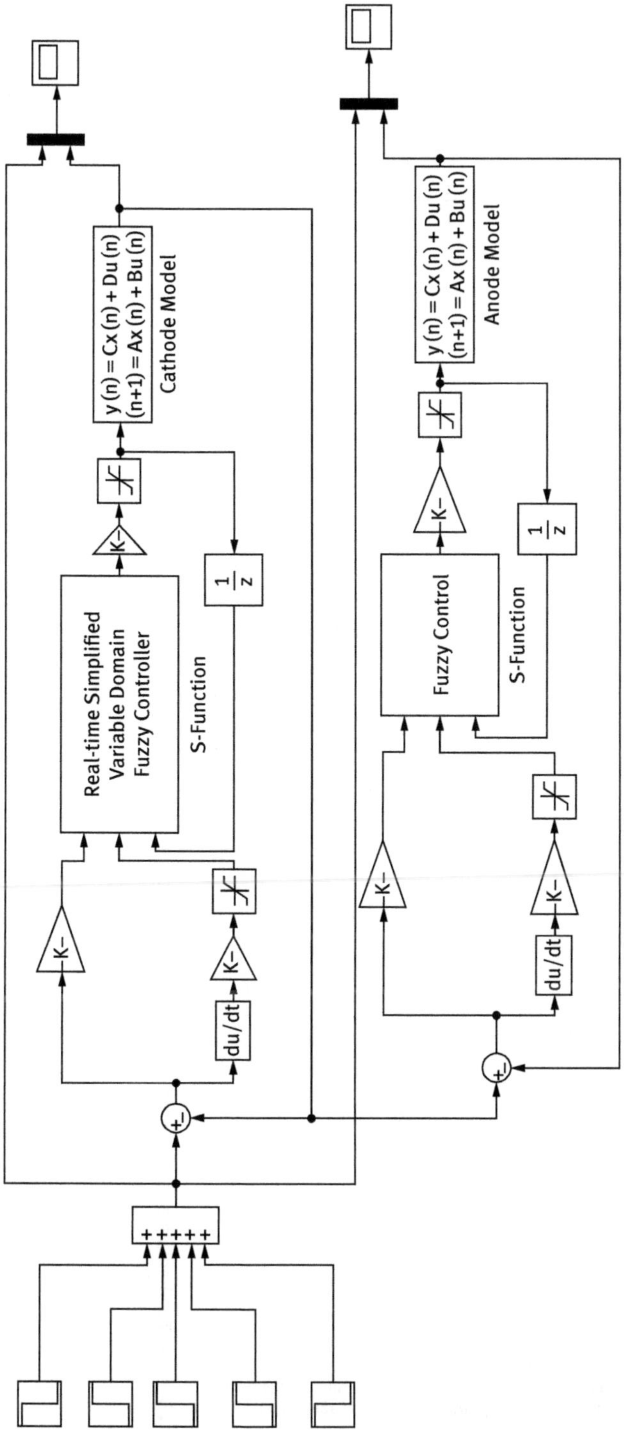

Figure 4.17: PEM fuel cell flow control system simulative diagram in Simulink.

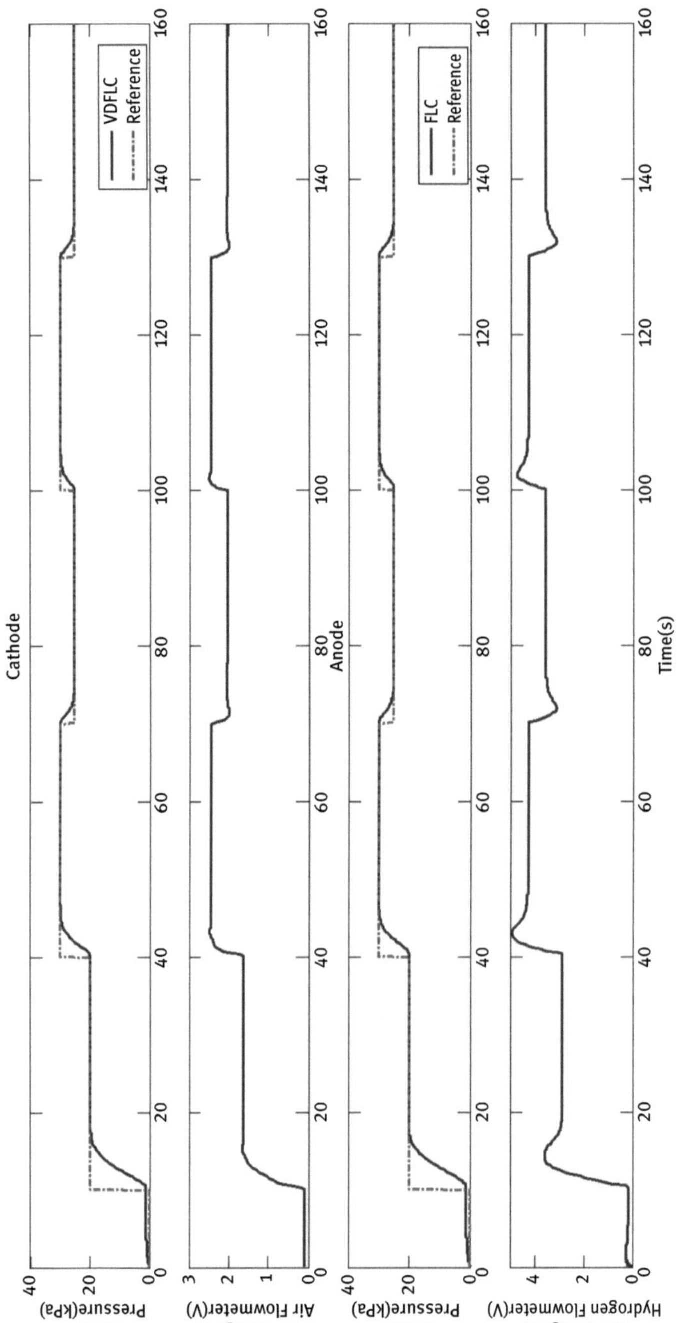

Figure 4.18: Simulative control performance for PEM fuel cell flow system.

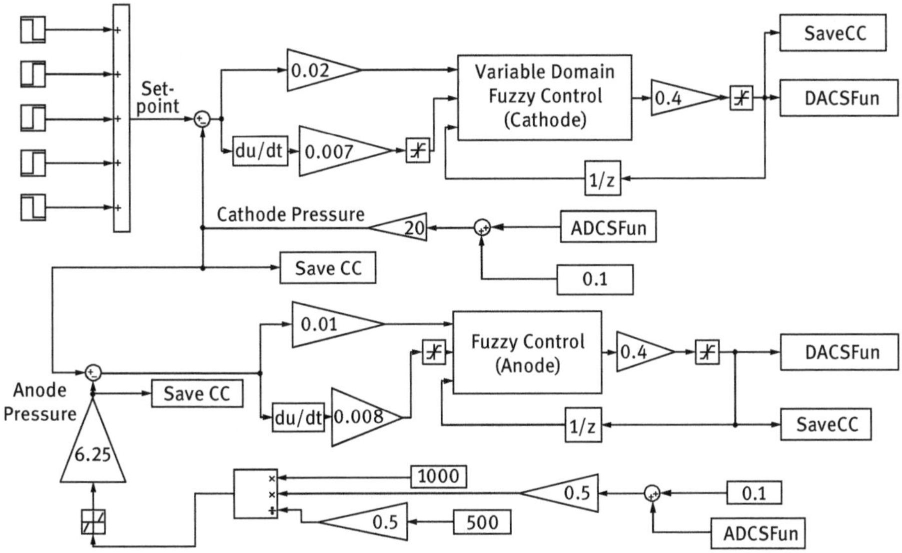

Figure 4.19: PEM fuel cell flow control system configuration diagram in Simulink.

is 0.1 second and the scaling gains K_e, K_{ec}, and K_u of the cathode are 0.02, 0.007, and 0.4, and the scaling gains for the anode are 0.01, 0.008 , and 0.4, respectively. $\tau_1 = \tau_2 = 0.1$, $\tau_3 = 0.7$. The 'ADCSFun' and 'DACSFun' modules are written using MATLAB's C-S functions. They are the interfaces between the controller, the external sensors, and the actuators. The 'SaveCC' is a storage module. S-function 'Variable Domain Fuzzy Control (Cathode)' is the core algorithm. In order to ensure that there is no steady-state error, the incremental algorithm is adopted. S-function 'Fuzzy Control (Anode)' is built in the same way expect the fixed domain. Using Real-time Workshop and cross-compilers provided by Simulink, the control configuration of Figure 4.18 can be translated into executable code to run in the ARM processor of the controller.

ii. Real-time fuzzy control

Figure 4.20 is the control performance of the real-time simplified VDFLC. It can be seen that the real-time simplified VDFLC method is more stable and more accurate. It can effectively reduce the PEM suffered stress and extend the life of the stack. Figure 4.21 is the corresponding control action. In order to verify the performance of the proposed real-time simplified VDFLC algorithm, three experiments are performed. In experiment 1, the method of VDFLC is used in the cathode, and the ordinary fuzzy control (FLC) is used in the anode. In experiment 2, FLC is used in both cathode and anode. Experiment 3 implements the proportional-integral control (PIC) for the cathode and anode. As shown in Figure 4.22, it is the control performance with empty load before the 150th second, while with a given load after the

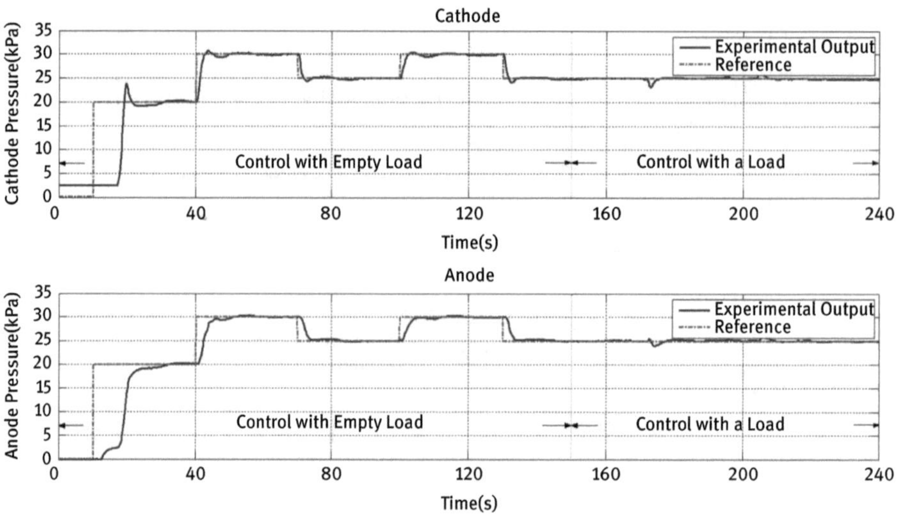

Figure 4.20: Real-time control for PEM fuel cell flow system (VDFC).

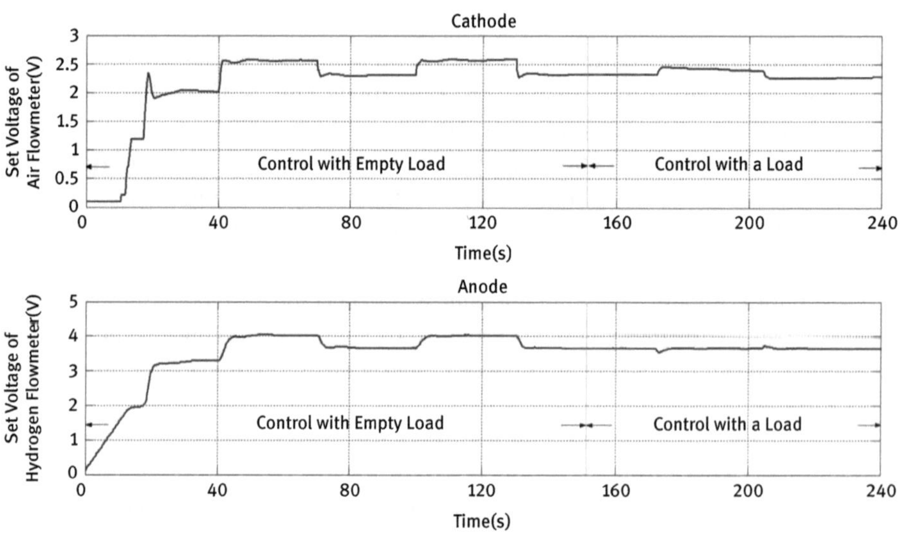

Figure 4.21: Controller outputs for the cathode and the anode (VDFLC).

150th second. Here, empty load means that the fuel cell has no current output but the fuel cell stacks are usually accompanied by pressure changes, a typical situation in the rising stage of fuel cells when it starts. To test this stage, we apply successive step responses to the system, as shown in Figure 4.22, the system changes from 0kPa to 20kPa in the 10th second and up to 30kPa in the 40th second, and then down to 25kPa

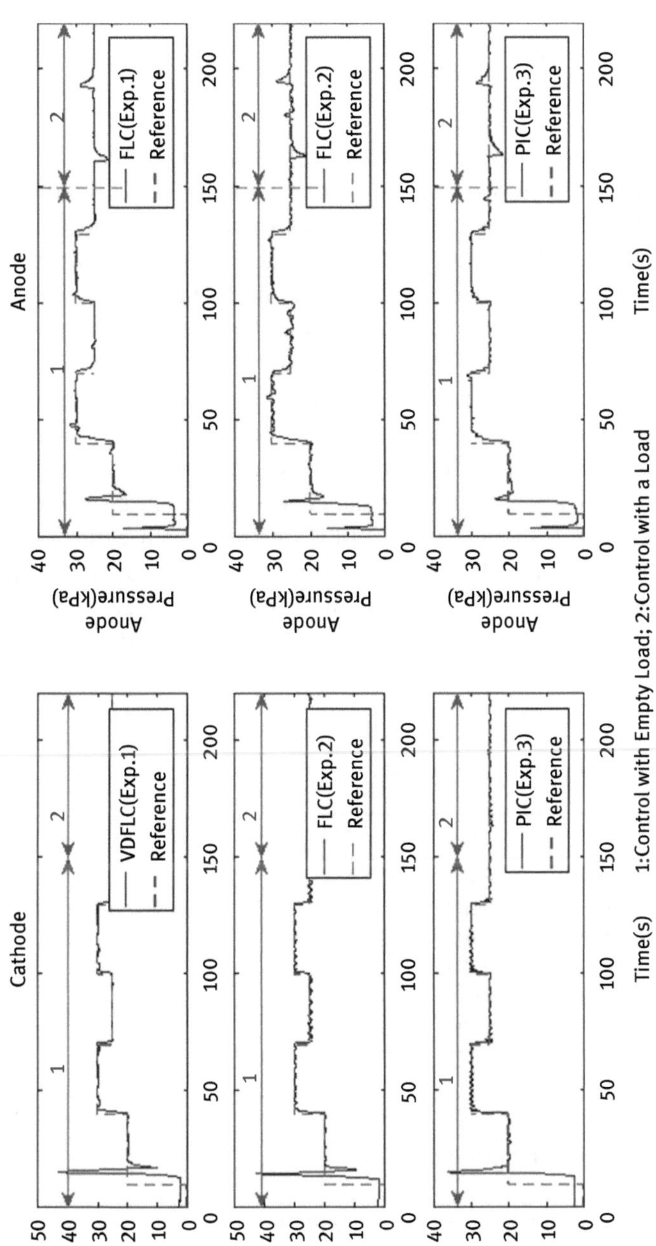

Figure 4.22: Control performance comparison between VDFLC, FLC, and PIC.

in the 70th second; in the 100th second, it rose to 30kPa again, and then down to 25kPa after the 130th second. In order to give an objective evaluation, we use the absolute integral error (IAE) as the evaluation index; the smaller the index, the better the control performance. Discrete forms of IAE can be defined as:

$$\mathrm{IAE} = \sum_{k=n_1}^{n_2} |e(k)| \cdot T_s$$

where T_s is the sample time, $T_s = 0.1$ second.

From Table 4.4, it can be seen that the PIC method is better than the FLC and VDFLC for data comparisons before the 22nd second, but the VDFLC method is better than the FLC and PIC after the 22nd second. This phenomenon proves that VDFLC method is better than FLC and PIC. Because the domain is rough at the beginning of the process, with the error changes, variable domain strategy begins to play a role, and control performance gets better and better to jointly meet the real-time and accuracy requirements. Through the expansion and contraction of the domain, precise control can be realized. Through the simplification of the reasoning process, real-time control can be realized. Moreover, the algorithm is achieved using C code, which will further speed up the implementation speed. In response to a load change of about 300 W, the data between the 150th second and the 220th second in Table 4.4 show that the VDFLC method has a better disturbances suppression ability than the FLC and PIC methods. Figure 4.23 is the output of the controller for the cathode and anode, and the output of the fuzzy controller is limited to 0–5V due to the physical limitation of the air and the hydrogen flowmeter.

Table 4.4: IAE comparison between VDFLC, FLC, and PIC.

Data No.	VDFLC	FLC	PIC
1–2200	181.9699	172.981	172.955
410–2200	46.0661	60.7189	46.5569
701–2200	30.2037	45.8208	33.9425
1001–2200	23.0585	36.4328	26.4217
1301–2200	15.1658	25.8445	17.9995
1501–2200	9.0069	18.2995	11.8981

4.3.3 Comparison of two control methods

In order to compare the control performance of SPM-based ASFPC and real-time simplified VDFLC method, a set of experiments were conducted on the actual fuel cell device. From the experimental results, we can see that the control performance of

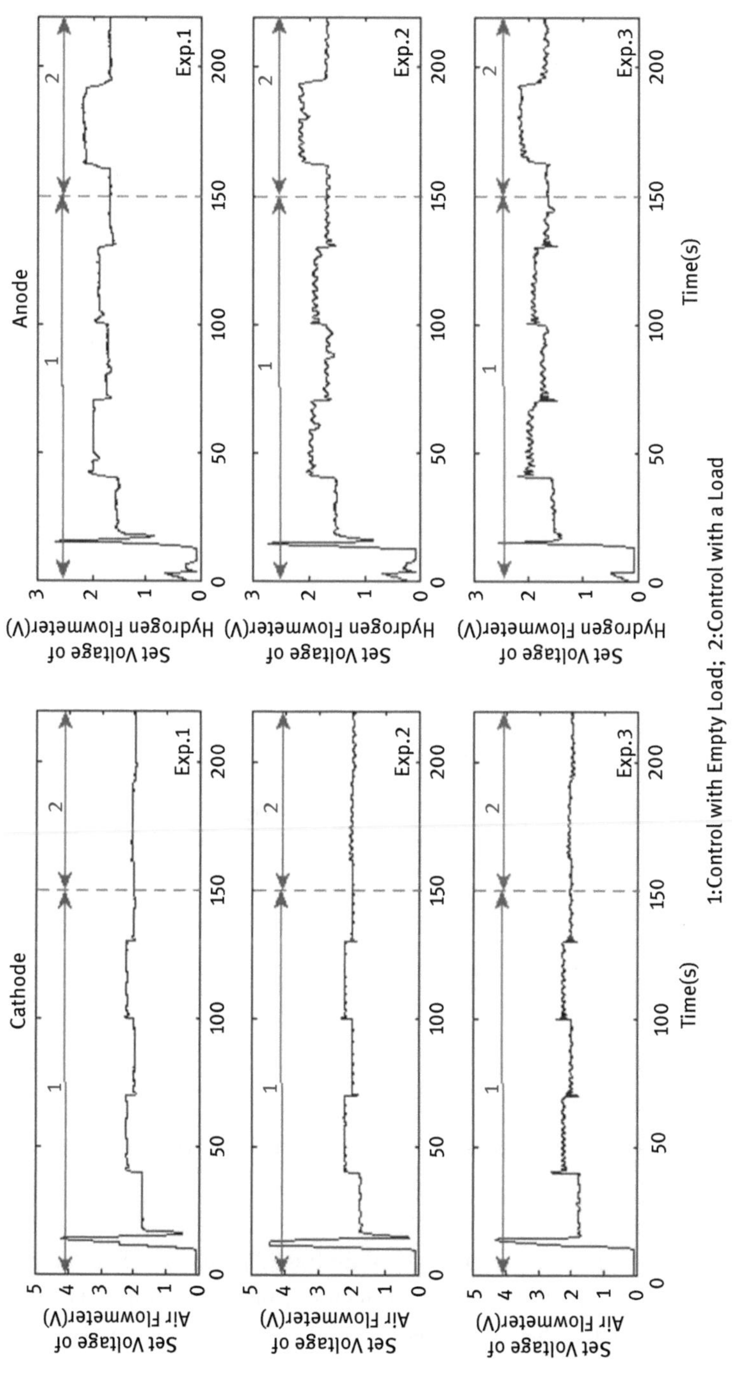

Figure 4.23: Control actions comparison between VDFLC, FLC, and PIC.

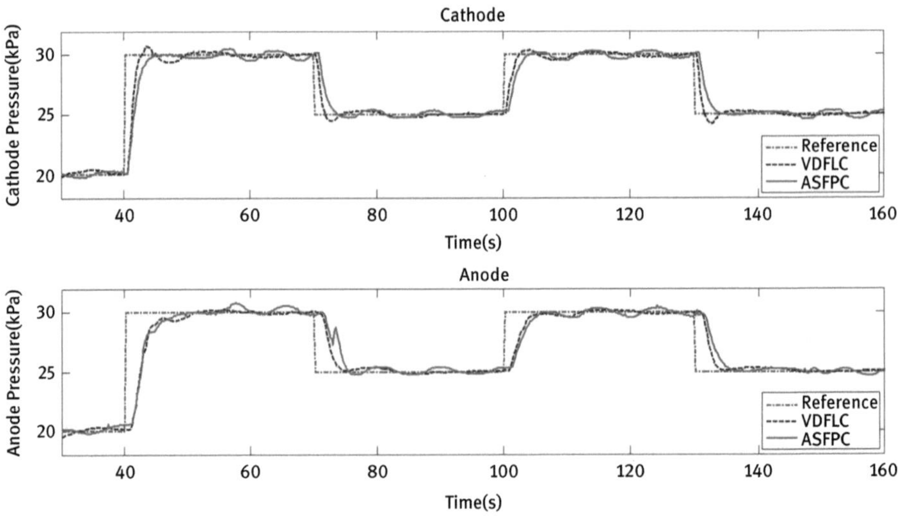

Figure 4.24: Control performance comparison between SPM-based ASFPC and VDFLC methods.

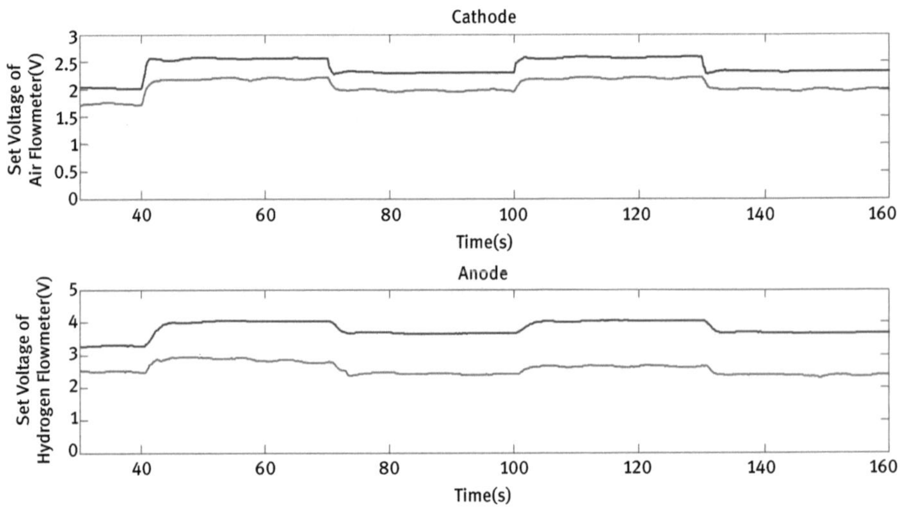

Figure 4.25: Control actions comparison between SPM-based ASFPC and VDFLC methods.

the real-time simplified VDFLC method is better than the SPM-based ASFPC method. Both the dynamic response and the steady-state accuracy have been improved. More importantly, this method has less computation and more suitable for real-time control environments.

4.4 Source codes

4.4.1 Joint estimation of parameters and states without considering noise information (M file)

```
/*M file: OfflineID_NoiseKnown============================
 * Abstract:
 * This M file is used to jointly estimate parameters and states with noise
   known.
 * A two-order system is used to simulate the algorithm.
 */
 Set_time=1000;
 nA=2;
 nD=0;
 nC=2;
 Yn=Set_time;
 u=idinput(Yn,'prbs',[4 1],[-15 15]);
 e=1*randn(1,Yn);
 y=zeros(1,Yn);
 a1=-0.2;
 a2=-0.24;
 b1=1;
 b2=2;
 c1=1.5;
 c2=0.75;

 for i=3:Yn
 y(i)=-a1*y(i-1)-a2*y(i-2)+b1*u(i-1)+b2*u(i-2)+e(i)+c1*e(i-1)+c2*e(i-2);
 end

 Out=[];
 E=[zeros(1,nA+nD);eye(nA+nD-1,nA+nD-1) zeros(nA+nD-1,1)];
 E0=[zeros(1,nA+nD-1) 1];
 H=[E0 zeros(1,2*(nA+nD))];
 C1=[zeros(1,nD),0,0];
 nC1=length(C1);
 s=zeros(3*(nA+nD),1);
 R=0;
 P1=10000*eye(3*(nA+nD),3*(nA+nD));
 S=[C1'*R; zeros(3*(nA+nD)-nC1,1)];
 Q1=[C1'*R*C1 zeros(nC1,3*(nA+nD)-nC1); zeros(3*(nA+nD)-nC1,3*(nA+nD))];
```

Note: For better understanding the source codes, some examples are presented. Please see Table 4.5–4.6 and Figure 4.26–4.30.

```
for k=1:Set_time
G1=[-y(k)*eye(nA+nD) u(k)*eye(nA+nD)];
F=[E G1;zeros(2*(nA+nD),nA+nD) eye(2*(nA+nD),2*(nA+nD))];
K1=(F*P1*H'+S)*inv(H*P1*H'+R);
P1=F*P1*F'+Q1-K1*(H*P1*H'+R)*K1';
s=F*s+K1*(y(k)-H*s)
end
```

Program running results:

Table 4.5: Simulative results for the joint estimation without considering noise.

Set_time	Lanbt	x_2	x_1	a2(−0.24)	a1(−0.2)	b2(1.0)	b1(2.0)	c2(0.75)	c1(1.5)	R
1000	–	−17.3336	36.5686	−0.2434	−0.2007	1.0016	2.0025	known	known	–

4.4.2 Joint estimation of parameters and states considering noise information (M file & C-Sfuntion)

1) M file

```
/*M file: OfflineID_NoiseUnknown==============================
 * Abstract:
 * This M file is used to jointly estimate parameters and states with noise
unknown.
 * A two-order system is used to simulate the algorithm.
 */

Set_time=1000;
nA=2;
nD=0;
nC=2;
Lanbt=8;
Yn=Set_time;
u=idinput(Yn,'prbs',[4 1],[-15 15]);
e=1*randn(1,Yn);
y=zeros(1,Yn);
a1=-0.2;
a2=-0.24;
b1=2;
b2=1;
c1=1.5;
```

```
c2=0.75;
for i=3:Yn
y(i)=-a1*y(i-1)-a2*y(i-2)+b1*u(i-1)+b2*u(i-2)+e(i)+c1*e(i-1)+c2*e(i-2);
end
E=[zeros(1,nA+nD);eye(nA+nD-1,nA+nD-1) zeros(nA+nD-1,1)];
E0=[zeros(1,nA+nD-1) 1];
H=[E0 zeros(1,3*(nA+nD))];
G0=[zeros(1,2*(nA+nD))];
s=zeros(4*(nA+nD),1);
R=1;
P1=10000*eye(4*(nA+nD),4*(nA+nD));

data=[];
for i=1:Set_time
e1=y(i)-H*s;
G1=[-y(i)*eye(nA+nD) u(i)*eye(nA+nD)];
F=[E G1 e1*eye(nA+nD);zeros(3*(nA+nD),nA+nD) eye(3*(nA+nD),3*(nA+nD))];
M=[E-s(3*(nA+nD)+1:4*(nA+nD))*E0         G1-s(3*(nA+nD)+1:4*(nA+nD))*G0
e1*eye(nA+nD); zeros(3*(nA+nD),nA+nD) eye(3*(nA+nD),3*(nA+nD))];
K1=M*P1*H'*inv(H*P1*H'+R);
P1=M*P1*M'-K1*(H*P1*H'+R)*K1';
R=R+Lanbt/(i+1)*(e1*e1'-R)
s=F*s+K1*(y(i)-H*s)
data=[data s];
end
```

Program running results:

Table 4.6: Simulative results for the joint estimation considering noise.

Set_time	Lanbt	x_2	x_1	a2	a1 (−0.2)	b2 (1.0)	b1 (2.0)	c2 (0.75)	c1 (1.5)	R
1000	8	−1.1462	7.7629	−0.2357	−0.2063	0.9891	2.0013	0.7116	1.4735	0.9993

2) C-Sfunction

```
/*C-Sfunction: OnlineID_SPM_NoiseUknown.c=========================
 * Abstract:
 *  This C-Sfunction is used to jointly estimate parameters and states
 *  with noise unknown.
 *  A two-order system is used to simulate the algorithm.
 */
```

```
#define S_FUNCTION_NAME OnlineID_SPM_NoiseUknown
#define S_FUNCTION_LEVEL 2

#include "simstruc.h"
#include "math.h"
#include "brmul.c"
#include "brtranspose.c"
#include "brinv.c"

/*=====================*
 * S-function methods *
 *=====================*/

/*Function:mdlInitializeSizes==========================================
 * Abstract:
 *  The sizes information is used by Simulink to determine the S-function
 *  block's characteristics (number of inputs, outputs, states, etc.).
 */
static int_T count;//Counting variable used to set initial value
#define sampletime(S) ssGetSFcnParam(S, 0)

static void mdlInitializeSizes(SimStruct *S)
{
    ssSetNumSFcnParams(S, 1);/* Number of expected parameters */
    if (ssGetNumSFcnParams(S) != ssGetSFcnParamsCount(S)) {
    /* Return if number of expected != number of actual parameters */
    return;
    }
    ssSetNumContStates(S, 0);
    ssSetNumDiscStates(S, 0);

    if (!ssSetNumInputPorts(S, 2)) return;  //Specify the number of input
ports that a block has.
    ssSetInputPortWidth(S, 0, 1);  //Specify the width of an input port.
    ssSetInputPortWidth(S, 1, 1);  //Specify the width of an input port.
    ssSetInputPortRequiredContiguous(S, 0, true);  /*direct input signal
access*/
    ssSetInputPortRequiredContiguous(S, 1, true);  /*direct input signal
access*/
   /*
    * Set direct feedthrough flag (1=yes, 0=no).
    * A port has direct feedthrough if the input is used in either
    * the mdlOutputs or mdlGetTimeOfNextVarHit functions.
    * See matlabroot/simulink/src/sfuntmpl_directfeed.txt.
```

```
    */
    ssSetInputPortDirectFeedThrough(S, 0, 1);//Specify the direct feed-
through status of a block's ports.
    ssSetInputPortDirectFeedThrough(S, 1, 1);//Specify the direct feed-
through status of a block's ports.
    //ssSetInputPortDirectFeedThrough(S, 2, 1);//Specify the direct feed-
through status of a block's ports.

    if (!ssSetNumOutputPorts(S, 1)) return;
    ssSetOutputPortWidth(S, 0, 8);

    ssSetNumSampleTimes(S, 1);     //Specify the number of sample times that
an S-Function block has.
    ssSetNumRWork(S, 0);
    ssSetNumIWork(S, 0);
    ssSetNumPWork(S, 0);
    ssSetNumModes(S, 0);
    ssSetNumNonsampledZCs(S, 0);

    ssSetOptions(S, 0);
}

/*Function: mdlInitializeSampleTimes==================================
 * Abstract:
 *  This function is used to specify the sample time(s) for your
 *  S-function. You must register the same number of sample times as
 *  specified in ssSetNumSampleTimes.
 */
static void mdlInitializeSampleTimes(SimStruct *S)
{
    const real_T *stime=mxGetPr(sampletime(S));
    ssSetSampleTime(S, 0, stime[0]);
    ssSetOffsetTime(S, 0, 0.0);
}

#define MDL_INITIALIZE_CONDITIONS/* Change to #undef to remove function
*/
#if defined(MDL_INITIALIZE_CONDITIONS)
/*Function: mdlInitializeConditions===================================
 * Abstract:
 *  In this function, you should initialize the continuous and discrete
 *  states for your S-function block. The initial states are placed
 *  in the state vector, ssGetContStates(S) or ssGetRealDiscStates(S).
 *  You can also perform any other initialization activities that your
```

```
 * S-function may require. Note, this routine will be called at the
 * start of simulation and if it is present in an enabled subsystem
 * configured to reset states, it will be call when the enabled subsystem
 * restarts execution to reset the states.
 */
 static void mdlInitializeConditions(SimStruct *S)
 {
 }
 #endif/* MDL_INITIALIZE_CONDITIONS */

 #define MDL_START/* Change to #undef to remove function */
 #if defined(MDL_START)
/*Function: mdlStart===============================================
 * Abstract:
 * This function is called once at start of model execution. If you
 * have states that should be initialized once, this is the place
 * to do it.
 */
 static void mdlStart(SimStruct *S)
 {
   count=0;
 }
 #endif/* MDL_START */

/*Function: mdlOutputs=============================================
 * Abstract:
 * In this function, you compute the outputs of your S-function
 * block. Generally outputs are placed in the output vector, ssGetY(S).
 */
 #define nA 2
 #define nD 0
 static void mdlOutputs(SimStruct *S, int_T tid)
 {
   const real_T *u = ssGetInputPortRealSignal(S,0);
   const real_T *y = ssGetInputPortRealSignal(S,1);
   real_T  *yout = (real_T *)ssGetOutputPortSignal(S,0);
   int_T i,j;
   const real_T  *stime=mxGetPr(sampletime(S));
   static real_T H[1][4*(nA+nD)],s[4*(nA+nD)][1];
   static real_T H1[4*(nA+nD)][1];
   static real_T P1[4*(nA+nD)][4*(nA+nD)],F[4*(nA+nD)][4*(nA+nD)];
   static real_T tmp_K1_1[4*(nA+nD)][4*(nA+nD)];
   static real_T tmp_K1_2[4*(nA+nD)][1];
```

```
  static real_T tmp_K1_3[1][4*(nA+nD)];
static real_T tmp_K1_4[1][1];
static real_T K1[4*(nA+nD)][1],tmp_s_2[1][1],tmp_K1_5[1][1];
  static real_T F1[4*(nA+nD)][4*(nA+nD)];
  static real_T K11[1][4*(nA+nD)],tmp_s_4[1][1];
  static real_T tmp_P1_1[4*(nA+nD)][1];
  static real_T tmp_P1_2[4*(nA+nD)][4*(nA+nD)];
  static real_T tmp_P1_3[4*(nA+nD)][4*(nA+nD)];
  static real_T tmp_s_1[4*(nA+nD)][1];
  static real_T tmp_s_3[4*(nA+nD)][1];
  static real_T R,e,tmp_K1_6[1][1],tmp_Hs[1][1];
  static real_T M[4*(nA+nD)][4*(nA+nD)],M1[4*(nA+nD)][4*(nA+nD)];

if (count==0)
{
for(i=0;i<4*(nA+nD);i++)
{
 if (i==(nA+nD-1))
    H[0][i]=1;
 else
    H[0][i]=0;
    s[i][0]=0;
}
for(i=0;i<4*(nA+nD);i++)
  for(j=0;j<4*(nA+nD);j++)
  {
    if(i==j)
     P1[i][j]=10000;
    else
     P1[i][j]=0;
  }
  R=1;
}
brmul(H,s,1,4*(nA+nD),1,tmp_Hs);
e=y[0]-tmp_Hs[0][0];
for(i=0;i<(nA+nD);i++)
for(j=0;j<(nA+nD);j++)
{
  if(j==(i-1))
     F[i][j]=1;
  else if(j==i)
  {
```

```
        F[i][j+nA+nD]=-y[0];
        F[i][j+2*(nA+nD)]=u[0];
        F[i][j+3*(nA+nD)]=e;
        F[i+nA+nD][j+nA+nD]=1;
        F[i+2*(nA+nD)][j+2*(nA+nD)]=1;
        F[i+3*(nA+nD)][j+3*(nA+nD)]=1;
      }
      else
        F[i][j]=0;
    }
    for(i=0;i<(nA+nD);i++)
    for(j=0;j<(nA+nD);j++)
    {
      if(j==(i-1))
        M[i][j]=1;
      else if(j==i)
      {
        M[i][j+nA+nD]=-y[0];
        M[i][j+2*(nA+nD)]=u[0];
        M[i][j+3*(nA+nD)]=e;
        M[i+nA+nD][j+nA+nD]=1;
        M[i+2*(nA+nD)][j+2*(nA+nD)]=1;
        M[i+3*(nA+nD)][j+3*(nA+nD)]=1;
      }
      else if(j==(nA+nD-1))
        M[i][j]=-s[3*(nA+nD)+i][0];
      else
        M[i][j]=0;
    }
    brtranspose(H,1,4*(nA+nD),H1);
    brmul(M,P1,4*(nA+nD),4*(nA+nD), 4*(nA+nD),tmp_K1_1);
    brmul(tmp_K1_1,H1,4*(nA+nD),4*(nA+nD),1,tmp_K1_2);
    brmul(H,P1,1, 4*(nA+nD), 4*(nA+nD),tmp_K1_3);
    brmul(tmp_K1_3,H1,1, 4*(nA+nD),1,tmp_K1_4);
    tmp_K1_5[0][0]=tmp_K1_4[0][0]+R;
    tmp_K1_6[0][0]=tmp_K1_5[0][0];
    brinv(tmp_K1_6,1);
    brmul(tmp_K1_2,tmp_K1_6, 4*(nA+nD),1,1,K1);
    brtranspose(M, 4*(nA+nD),4*(nA+nD),M1);
    brtranspose(K1, 4*(nA+nD),1,K11);
    brmul(K1,tmp_K1_5, 4*(nA+nD),1,1,tmp_P1_1);
    brmul(tmp_P1_1,K11, 4*(nA+nD),1,4*(nA+nD),tmp_P1_2);
```

```
brmul(tmp_K1_1, M1,4*(nA+nD),4*(nA+nD), 4*(nA+nD),tmp_P1_3);

for(i=0;i<4*(nA+nD);i++)
   for(j=0;j<4*(nA+nD);j++)
P1[i][j]=tmp_P1_3[i][j]-tmp_P1_2[i][j];

brmul(F, s,4*(nA+nD), 4*(nA+nD),1,tmp_s_1);
brmul(H, s,1, 4*(nA+nD),1,tmp_s_2);
tmp_s_4[0][0]=y[0]-tmp_s_2[0][0];
brmul(K1, tmp_s_4,4*(nA+nD),1,1,tmp_s_3);

for (i=0;i<4*(nA+nD);i++)
s[i][0]=tmp_s_1[i][0]+tmp_s_3[i][0];

R=R+1/(count+1)*(e*e-R);

for(i=0;i<4*(nA+nD);i++)
yout[i]=s[i][0];

count=count+1;
}

#define MDL_UPDATE/* Change to #undef to remove function */
#if defined(MDL_UPDATE)
/*Function: mdlUpdate=============================================
* Abstract:
*  This function is called once for every major integration time step.
*  Discrete states are typically updated here, but this function is useful
*  for performing any tasks that should only take place once per
*  integration step.
*/
static void mdlUpdate(SimStruct *S, int_T tid)
{
}
#endif/* MDL_UPDATE */

#define MDL_DERIVATIVES/* Change to #undef to remove function */
#if defined(MDL_DERIVATIVES)
/Function: mdlDerivatives=========================================
* Abstract:
*  In this function, you compute the S-function block's derivatives.
*  The derivatives are placed in the derivative vector, ssGetdX(S).
*/
static void mdlDerivatives(SimStruct *S)
```

```
{
}
#endif/* MDL_DERIVATIVES */

/*Function: mdlTerminate========================================
 * Abstract:
 *  In this function, you should perform any actions that are necessary
 *  at the termination of a simulation. For example, if memory was
 *  allocated in mdlStart, this is the place to free it.
 */
 static void mdlTerminate(SimStruct *S)
 {
 }

/*=========================================================*
 * See sfuntmpl_doc.c for the optional S-function methods *
 *=========================================================*/

/*=============================*
 * Required S-function trailer *
 *=============================*/

 #ifdef MATLAB_MEX_FILE/* Is this file being compiled as a MEX-file? */
 #include "simulink.c"  /* MEX-file interface mechanism */
 #else
 #include "cg_sfun.h"  /* Code generation registration function */
 endif

/*C file: BRINV.c ==========================================
 * Abstract:
 *  This C file is programmed for Matrix Inversion.
 */

 #include "stdlib.h"
 #include "math.h"
 #include "stdio.h"
 int brinv(a,n)
 int n;
 double a[];
 {int *is,*js,i,j,k,l,u,v;
 double d,p;
 is=malloc(n*sizeof(int));
 js=malloc(n*sizeof(int));
```

```
for (k=0; k≤n-1; k++)
 {d=0.0;
  for (i=k; i≤n-1; i++)
  for (j=k; j≤n-1; j++)
   {l=i*n+j; p=fabs(a[l]);
   if (p>d) {d=p; is[k]=i; js[k]=j;}
   }
if (d+1.0==1.0)
 {free(is); free(js); printf("err**not inv\n");
  return(0);
 }
 if (is[k]!=k)
  for (j=0; j≤n-1; j++)
   {u=k*n+j; v=is[k]*n+j;
    p=a[u]; a[u]=a[v]; a[v]=p;
   }
 if (js[k]!=k)
  for (i=0; i≤n-1; i++)
   {u=i*n+k; v=i*n+js[k];
    p=a[u]; a[u]=a[v]; a[v]=p;
   }
 l=k*n+k;
 a[l]=1.0/a[l];
 for (j=0; j≤n-1; j++)
  if (j!=k)
   {u=k*n+j; a[u]=a[u]*a[l];}
 for (i=0; i≤n-1; i++)
  if (i!=k)
    for (j=0; j≤n-1; j++)
     if (j!=k)
      {u=i*n+j;
       a[u]=a[u]-a[i*n+k]*a[k*n+j];
      }
 for (i=0; i≤n-1; i++)
  if (i!=k)
   {u=i*n+k; a[u]=-a[u]*a[l];}
 }
 for (k=n-1; k>=0; k-)
  {if (js[k]!=k)
   for (j=0; j≤n-1; j++)
    {u=k*n+j; v=js[k]*n+j;
     p=a[u]; a[u]=a[v]; a[v]=p;
```

```
    }
  if (is[k]!=k)
   for (i=0; i≤n-1; i++)
    {u=i*n+k; v=i*n+is[k];
     p=a[u]; a[u]=a[v]; a[v]=p;
    }
  }
  free(is); free(js);
  return(1);
  }

/*C file: BRMUL.c ============================================
 * Abstract:
 * This C file is programmed for Matrix Multiplication.
 */

 int brmul(a,b,m,n,k,c)
 int m,n,k;
 double a[],b[],c[];
 {int i,j,l,u;
 for (i=0; i≤m-1; i++)
 for (j=0; j≤k-1; j++)
   {u=i*k+j; c[u]=0.0;
    for (l=0; l≤n-1; l++)
      c[u]=c[u]+a[i*n+l]*b[l*k+j];
   }
  return(1);
 }

/*C file: BRTRANSPOSE.c ========================================
 * Abstract:
 *  This C file is programmed for Matrix Transpose.
 */
 int brtranspose(a,m,n,b)
  int m,n;
  double a[],b[];
  {int i,j;
    for(i=0;i<n;i++)
    for(j=0;j<m;j++)
    b[i*m+j]=a[j*n+i];
    return(1);
 }
```

4.4.3 ASFPC algorithm based on SPM

1) Simulink diagram of a sample program for ASFPC_SPM

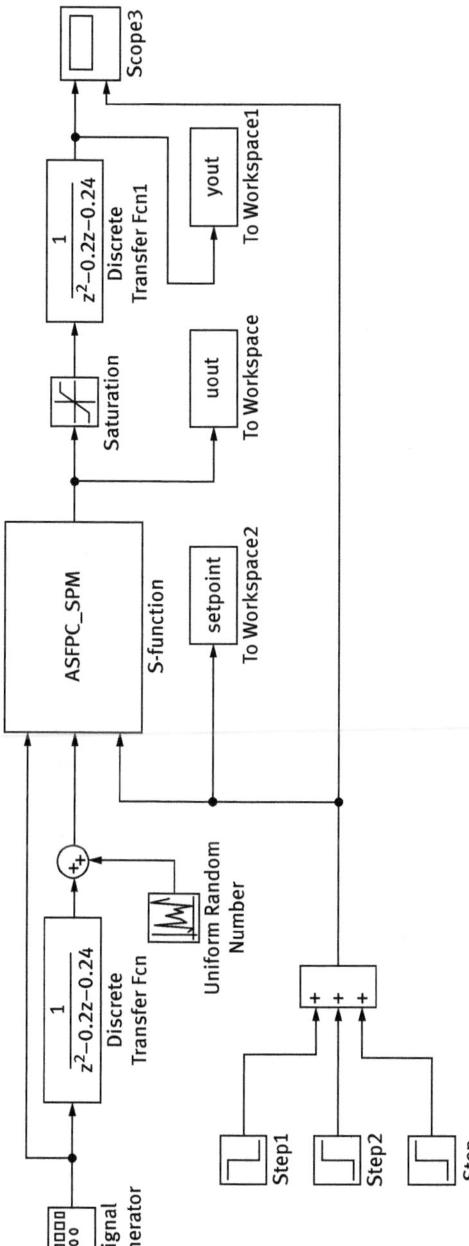

Figure 4.26: Simulink diagram of SPM-based ASFPC algorithm for a two-order system.

2) C-Sfunction for ASFPC_SPM

```
/*C-Sfunction: ASFPC_SPM.c=========================================
 * Abstract:
 *  This C-Sfunction is used to simulate the SPM-based ASFPC algorithm.
 */

 #define S_FUNCTION_NAME ASFPC_SPM
 #define S_FUNCTION_LEVEL 2

/*
 * Need to include simstruc.h for the definition of the SimStruct and
 * its associated macro definitions.
 */
 #include "simstruc.h"
 #include "math.h"
 #include "brmul.c"
 #include "brtranspose.c"
 #include "brinv.c"
 #include "count_Sp.c"

/* Error handling
 *
 * You should use the following technique to report errors encountered within
 * an S-function:
 *
 *    ssSetErrorStatus(S,"Error encountered due to ...");
 *    return;
 *
 * Note that the 2nd argument to ssSetErrorStatus must be persistent
memory.
 * It cannot be a local variable. For example the following will cause
 * unpredictable errors:
 *
 *    mdlOutputs()
 *    {
 *    char msg[256]; {ILLEGAL: to fix use "static char msg[256];"}
 *    sprintf(msg,"Error due to %s", string);
 *    ssSetErrorStatus(S,msg);
 *    return;
 * }
 *
 * See matlabroot/simulink/src/sfuntmpl_doc.c for more details.
 */
```

```
/*====================*
 * S-function methods *
 *====================*/

/*Function: mdlInitializeSizes ========================================
 * Abstract:
 *  The sizes information is used by Simulink to determine the S-function
 *  block's characteristics (number of inputs, outputs, states, etc.).
 */
 static int_T count; //Counting variable used to set initial value
 #define sampletime(S) ssGetSFcnParam(S, 0)

 static void mdlInitializeSizes(SimStruct *S)
 {
   /* See sfuntmpl_doc.c for more details on the macros below */

   ssSetNumSFcnParams(S, 1);/* Number of expected parameters */
   if (ssGetNumSFcnParams(S) != ssGetSFcnParamsCount(S)) {
      /* Return if number of expected != number of actual parameters */
      return;
 }
  ssSetNumContStates(S, 0);
  ssSetNumDiscStates(S, 0);
  //if (!ssSetNumInputPorts(S, 1)) return;//Specify the number of input
ports that a block has.
  if (!ssSetNumInputPorts(S, 3)) return; //Specify the number of input
ports that a block has.
  ssSetInputPortWidth(S, 0, 1); //Specify the width of an input port.
  ssSetInputPortWidth(S, 1, 1); //Specify the width of an input port.
  ssSetInputPortWidth(S, 2, 1); //Specify the width of an input port.
  ssSetInputPortRequiredContiguous(S, 0, true);/*direct input signal
access*/
  ssSetInputPortRequiredContiguous(S, 1, true);/*direct input signal access*/
  ssSetInputPortRequiredContiguous(S, 2, true);/*direct input signal
access*/
 /*

 * Set direct feedthrough flag (1=yes, 0=no).
 * A port has direct feedthrough if the input is used in either
 * the mdlOutputs or mdlGetTimeOfNextVarHit functions.
 * See matlabroot/simulink/src/sfuntmpl_directfeed.txt.
 */
 ssSetInputPortDirectFeedThrough(S, 0, 1);//Specify the direct feedthrough
status of a block's ports.
```

```
    ssSetInputPortDirectFeedThrough(S, 1, 1);//Specify the direct feedthrough
status of a block's ports.
    ssSetInputPortDirectFeedThrough(S, 2, 1);//Specify the direct feedthrough
status of a block's ports.

    if (!ssSetNumOutputPorts(S, 1)) return;
    ssSetOutputPortWidth(S, 0, 1);

    ssSetNumSampleTimes(S, 1);//Specify the number of sample times that an
S-Function block has.
    ssSetNumRWork(S, 0);
    ssSetNumIWork(S, 0);
    ssSetNumPWork(S, 0);
    ssSetNumModes(S, 0);
    ssSetNumNonsampledZCs(S, 0);

    ssSetOptions(S, 0);
  }

  /*Function: mdlInitializeSampleTimes ==================================
   * Abstract:
   *  This function is used to specify the sample time(s) for your
   *  S-function. You must register the same number of sample times as
   *  specified in ssSetNumSampleTimes.
   */
   static void mdlInitializeSampleTimes(SimStruct *S)
   {
    const real_T *stime=mxGetPr(sampletime(S));
    ssSetSampleTime(S, 0, stime[0]);
    ssSetOffsetTime(S, 0, 0.0);
   }

   #define MDL_INITIALIZE_CONDITIONS/* Change to #undef to remove function */
   #if defined(MDL_INITIALIZE_CONDITIONS)
   /*Function: mdlInitializeConditions ================================
    * Abstract:
    *  In this function, you should initialize the continuous and discrete
    *  states for your S-function block. The initial states are placed
    *  in the state vector, ssGetContStates(S) or ssGetRealDiscStates(S).
    *  You can also perform any other initialization activities that your
    *  S-function may require. Note, this routine will be called at the
    *  start of simulation and if it is present in an enabled subsystem
    *  configured to reset states, it will be call when the enabled subsystem
```

```
 * restarts execution to reset the states.
 */
static void mdlInitializeConditions(SimStruct *S)
{
}
#endif/* MDL_INITIALIZE_CONDITIONS */

#define MDL_START/* Change to #undef to remove function */
#if defined(MDL_START)
/*Function: mdlStart ===========================================
 * Abstract:
 *  This function is called once at start of model execution. If you
 *  have states that should be initialized once, this is the place
 *  to do it.
 */
static void mdlStart(SimStruct *S)
{
 count=0;
}
#endif/* MDL_START */

/*Function: mdlOutputs =========================================
 * Abstract:
 *  In this function, you compute the outputs of your S-function
 *  block. Generally outputs are placed in the output vector, ssGetY(S).
 */
#define nA 2
#define nD 0
#define P 10
#define L 20000
static void mdlOutputs(SimStruct *S, int_T tid)
{
  const real_T *u = ssGetInputPortRealSignal(S,0);
  const real_T *y = ssGetInputPortRealSignal(S,1);
  const real_T *Ys = ssGetInputPortRealSignal(S,2);

 real_T  *yout = (real_T *)ssGetOutputPortSignal(S,0);
 int_T i,j,k,m;

 const real_T *stime=mxGetPr(sampletime(S));
 static real_T H[1][4*(nA+nD)],s[4*(nA+nD)][1];
 static real_T H1[4*(nA+nD)][1];
```

```
static real_T P1[4*(nA+nD)][4*(nA+nD)],F[4*(nA+nD)][4*(nA+nD)];
static real_T tmp_K1_1[4*(nA+nD)][4*(nA+nD)];
static real_T tmp_K1_2[4*(nA+nD)][1];
static real_T tmp_K1_3[1][4*(nA+nD)];
static real_T tmp_K1_4[1][1],K1[4*(nA+nD)][1],tmp_s_2[1][1],tmp_K1_5[1][1];
static real_T F1[4*(nA+nD)][4*(nA+nD)];
static real_T K11[1][4*(nA+nD)],tmp_s_4[1][1];
static real_T tmp_P1_1[4*(nA+nD)][1];
static real_T tmp_P1_2[4*(nA+nD)][4*(nA+nD)];
static real_T tmp_P1_3[4*(nA+nD)][4*(nA+nD)];
static real_T tmp_s_1[4*(nA+nD)][1];
static real_T tmp_s_3[4*(nA+nD)][1],SP;
static real_T R,e,tmp_K1_6[1][1],tmp_Hs[1][1];
static real_T M[4*(nA+nD)][4*(nA+nD)],M1[4*(nA+nD)][4*(nA+nD)];

static real_T A[nA][nA],B[nA][1],C[1][nA],Q;
static real_T tmp_1[nA][nA],tmp_2[nA][nA],K[1][nA];
static real_T Fu[1][1][P],Fuhi[1][1][P];
static real_T X[nA][L],Y[1][L],tmp_X[nA][1],tmp_Y[1][1];
static real_T KX[1][1],FU_U;
static real_T Xk_tmp1[nA][1],KX_tmp1[1][nA],KX_tmp2[nA][1];
static real_T Bata_X[nA][nA],delt_u_tmp,delt_U[1][L],U[1][L];
static real_T tmp_X1[nA][1],tmp_X2[nA][1],tmp_U[1][1];

////////////////Initialization////////////////////////
if (count==0)
{
  Q=1;
  for(i=0;i<nA;i++)
    for(j=0;j<nA;j++)
 {
 if(j==i)
   Bata_X[i][j]=1;
 else
   Bata_X[i][j]=0;
 }
for(i=0;i<4*(nA+nD);i++)
{
 if (i==(nA+nD-1))
  H[0][i]=1;
else
  H[0][i]=0;
```

```
  s[i][0]=0;
 }
 for(i=0;i<4*(nA+nD);i++)
  for(j=0;j<4*(nA+nD);j++)
  {
    if(i==j)
      P1[i][j]=10000;
     else
      P1[i][j]=0;
  }
 for(i=0;i<nA;i++)
    for(j=0;j<L;j++)
    {
      X[i][j]=0;
      delt_U[0][j]=0;
      U[0][j]=0;
    }
R=1;
}
 count=count+1;
/////////////////////////////////////////////////////////

/////////////////////////////////////////////////////////
 //Calculate"F=[E G1;zeros(2*(nA+nD),nA+nD) eye(2*(nA+nD),2*(nA+nD))]"//
 //where,"G1=[-y1(k)*eye(nA+nD) u1(k)*eye(nA+nD)]"//
 brmul(H,s,1,4*(nA+nD),1,tmp_Hs);
 e=y[0]-tmp_Hs[0][0];
 for(i=0;i<(nA+nD);i++)
 for(j=0;j<(nA+nD);j++)
 {
   if(j==(i-1))
     F[i][j]=1;
   else if(j==i)
 {
     F[i][j+nA+nD]=-y[0];
     F[i][j+2*(nA+nD)]=u[0];
     F[i][j+3*(nA+nD)]=e;
     F[i+nA+nD][j+nA+nD]=1;
     F[i+2*(nA+nD)][j+2*(nA+nD)]=1;
     F[i+3*(nA+nD)][j+3*(nA+nD)]=1;
  }
  else
```

```
    F[i][j]=0;
  }
//////////////////////////////////////////////////////

////////////////////calculate "M"//////////////
  for(i=0;i<(nA+nD);i++)
  for(j=0;j<(nA+nD);j++)
  {
    if(j==(i-1))
      M[i][j]=1;
    else if(j==i)
    {
      M[i][j+nA+nD]=-y[0];
      M[i][j+2*(nA+nD)]=u[0];
      M[i][j+3*(nA+nD)]=e;
      M[i+nA+nD][j+nA+nD]=1;
      M[i+2*(nA+nD)][j+2*(nA+nD)]=1;
      M[i+3*(nA+nD)][j+3*(nA+nD)]=1;
    }
   else if(j==(nA+nD-1))
      M[i][j]=-s[3*(nA+nD)+i][0];
   else
      M[i][j]=0;
  }
//////////////////////////////////////////////////////

////////////////Calculate "K1=(F*P1*H'+S)*inv(H*P1*H'+R)"//
  brtranspose(H,1,4*(nA+nD),H1);
  brmul(M,P1,4*(nA+nD),4*(nA+nD), 4*(nA+nD),tmp_K1_1);
  brmul(tmp_K1_1,H1,4*(nA+nD),4*(nA+nD),1,tmp_K1_2);
  brmul(H,P1,1, 4*(nA+nD), 4*(nA+nD),tmp_K1_3);
  brmul(tmp_K1_3,H1,1, 4*(nA+nD),1,tmp_K1_4);
  tmp_K1_5[0][0]=tmp_K1_4[0][0]+R;
  tmp_K1_6[0][0]=tmp_K1_5[0][0];
  brinv(tmp_K1_6,1);
  brmul(tmp_K1_2,tmp_K1_6, 4*(nA+nD),1,1,K1);
//////////////////////////////////////////////////////

////////////////Calculate "P1=F*P1*F'+Q1-K1*(H*P1*H'+R)*K1'"//
  brtranspose(M, 4*(nA+nD),4*(nA+nD),M1);
  brtranspose(K1, 4*(nA+nD),1,K11);
  brmul(K1,tmp_K1_5, 4*(nA+nD),1,1,tmp_P1_1);
```

```
 brmul(tmp_P1_1,K11, 4*(nA+nD),1,4*(nA+nD),tmp_P1_2);
 brmul(tmp_K1_1, M1,4*(nA+nD),4*(nA+nD), 4*(nA+nD),tmp_P1_3);
 for(i=0;i<4*(nA+nD);i++)
    for(j=0;j<4*(nA+nD);j++)
 P1[i][j]=tmp_P1_3[i][j]-tmp_P1_2[i][j];
///////////////////////////////////////////////////////

///////////////////////Calculate “s=F*s+K1*(y1(k)-H*s)”//
 brmul(F, s,4*(nA+nD), 4*(nA+nD),1,tmp_s_1);
 brmul(H, s,1, 4*(nA+nD),1,tmp_s_2);
 tmp_s_4[0][0]=y[0]-tmp_s_2[0][0];
 brmul(K1, tmp_s_4,4*(nA+nD),1,1,tmp_s_3);
 for (i=0;i<4*(nA+nD);i++)
 s[i][0]=tmp_s_1[i][0]+tmp_s_3[i][0];
 R=R+1/(count+1)*(e*e-R);
///////////////////////////////////////////////////////

////////////Construct Matrix A,B,C///////////////////
 for(i=0;i<nA;i++)
 for(j=0;j<nA;j++)
 {
  if(j==nA-1)
    A[i][j]=-s[i+nA][0];
  else if(i==j+1)
    A[i][j]=1;
  else
    A[i][j]=0;
 }
 for(i=0;i<nA;i++)
 {B[i][0]=s[i+2*(nA)][0];}
 for(i=0;i<nA;i++)
 {
  if (i==nA-1)
    C[0][i]=1;
  else
    C[0][i]=0;
 }
////////////////////////////////////////////////////

////////////Calculate Matrix S(P)///////////////////
 SP=count_Sp(A,B,C,P);
/////////////////////////////////////////////////////
```

```
/////////Calculate Matrix K////////////////////////
 for(i=0;i<nA;i++)
 for(j=0;j<nA;j++)
 {
   if(j==i)
     tmp_1[i][j]=1;
   else
     tmp_1[i][j]=0;
   tmp_2[i][j]=0;
 }
 for(i=0;i<P;i++)
 {
   brmul(tmp_1,A,nA,nA,nA,tmp_2);
   for(j=0;j<nA;j++)
     for(k=0;k<nA;k++)
   tmp_1[j][k]=tmp_2[j][k];
 }
 brmul(C,tmp_2,1,nA,nA,K);
/////////////////////////////////////////////////////

/////////Calculate Control Feedback Parameter Fu////////
 for(i=0;i<P;i++)
   {Fu[0][0][i]=0;Fuhi[0][0][i]=0;}
 for (m=0;m<P;m++)
   Fuhi[0][0][m]=count_Sp(A,B,C,P)-count_Sp(A,B,C,m+1);
 for(i=0;i<P;i++)
   Fu[0][0][i]=Fu[0][0][i]+Fuhi[0][0][i];
///////////////////////////////////////////////////

////////////////Calculate Y////////////////////////////
 for (i=0;i<nA;i++)
   tmp_X[i][0]=X[i][count];
 brmul(C,tmp_X,1,nA,1,tmp_Y);
 Y[0][count]=tmp_Y[0][0];
 KX[0][0]=0;
 FU_U=0;
/////////////////////////////////////////////////////

/////////////Calculate State Feedback Matrix KX////////
 if (count≤P)
 {
   for(i=0;i<nA;i++)
```

```
      Xk_tmp1[i][0]=0;
  }
  else
  {
  for(i=0;i<nA;i++)
    Xk_tmp1[i][0]=X[i][count-P];
  }
  brmul(K,Bata_X,1,nA,nA,KX_tmp1);
  for(i=0;i<nA;i++)
    KX_tmp2[i][0]=tmp_X[i][0]-Xk_tmp1[i][0];
  brmul(KX_tmp1,KX_tmp2,1,nA,1,KX);
////////////////////////////////////////////////////////////

////////Calculate Dynamic Input Feedback Matrix FU_U////
  for(i=1;i≤P;i++)
  {
    if(count≤i)
      delt_u_tmp=0;
    else
      delt_u_tmp=delt_U[0][count-i];
    FU_U=FU_U+Fu[0][0][i-1]*delt_u_tmp;
  }
////////////////////////////////////////////////////////////

/////////Calculate delt_U//////////////////////////////
  if(fabs(SP)>0)
    delt_U[0][count]=1/(SP*Q*SP)*SP*Q*(Ys[0]-tmp_Y[0][0]-KX[0][0]-FU_U);
  else
    delt_U[0][count]=0;
/////////////////////////////////////////////////////////////

/////////Simulation of Process////////////////////////
   if(count≤1)
   U[0][count]=delt_U[0][count];
   else
   U[0][count]=delt_U[0][count]+U[0][count-1];
   tmp_U[0][0]=U[0][count];
   brmul(A,tmp_X,nA,nA,1,tmp_X1);
   brmul(B,tmp_U,nA,1,1,tmp_X2);
   for(i=0;i<nA;i++)
   X[i][count+1]=tmp_X1[i][0]+tmp_X2[i][0];
   yout[0]=U[0][count];
```

```
}

#define MDL_UPDATE/* Change to #undef to remove function */
#if defined(MDL_UPDATE)
/*Function: mdlUpdate ============================================
* Abstract:
*  This function is called once for every major integration time step.
*  Discrete states are typically updated here, but this function is useful
*  for performing any tasks that should only take place once per
*  integration step.
*/
static void mdlUpdate(SimStruct *S, int_T tid)
{
}
#endif/* MDL_UPDATE */

#define MDL_DERIVATIVES/* Change to #undef to remove function */
#if defined(MDL_DERIVATIVES)
/Function: mdlDerivatives ==========================================
* Abstract:
*  In this function, you compute the S-function block's derivatives.
*  The derivatives are placed in the derivative vector, ssGetdX(S).
*/
static void mdlDerivatives(SimStruct *S)
{
}
#endif/* MDL_DERIVATIVES */

/*Function: mdlTerminate ============================================
* Abstract:
*  In this function, you should perform any actions that are necessary
*  at the termination of a simulation. For example, if memory was
*  allocated in mdlStart, this is the place to free it.
*/
static void mdlTerminate(SimStruct *S)
{
}

/*======================================================*
* See sfuntmpl_doc.c for the optional S-function methods *
*======================================================*/
```

```
/*===============================*
 * Required S-function trailer *
 *===============================*/

#ifdef MATLAB_MEX_FILE /* Is this file being compiled as a MEX-file? */
#include "simulink.c" /* MEX-file interface mechanism */
#else
#include "cg_sfun.h" /* Code generation registration function */
#endif
```

3) Program running result

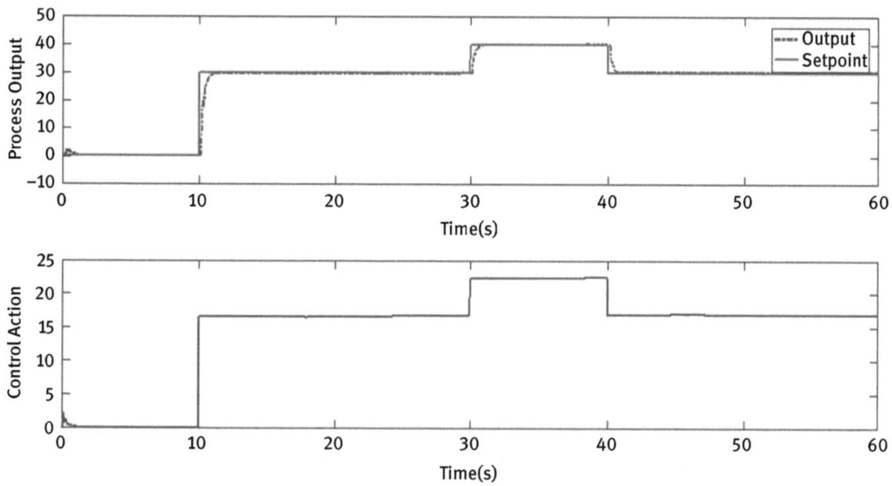

Figure 4.27: Simulative result of SPM-based ASFPC algorithm for a two-order system.

4.4.4 Real-time simplified VDFLC algorithm

1) Simulink diagram of a sample program for real-time simplified VDFLC

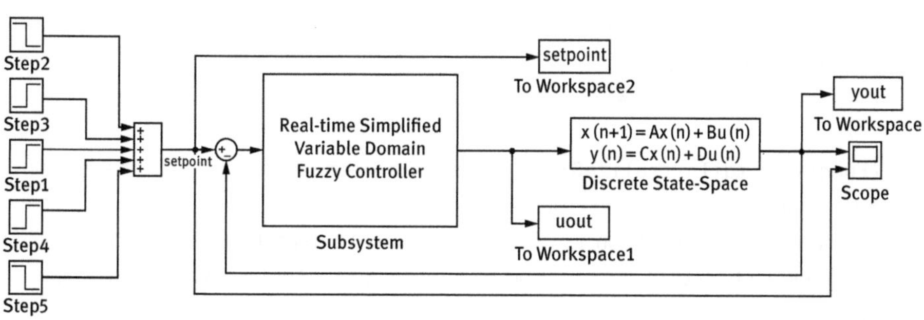

Figure 4.28: Simulink diagram of real-time simplified VDFLC algorithm.

2) Simulink diagram of the real-time simplified variable domain fuzzy controller subsystem

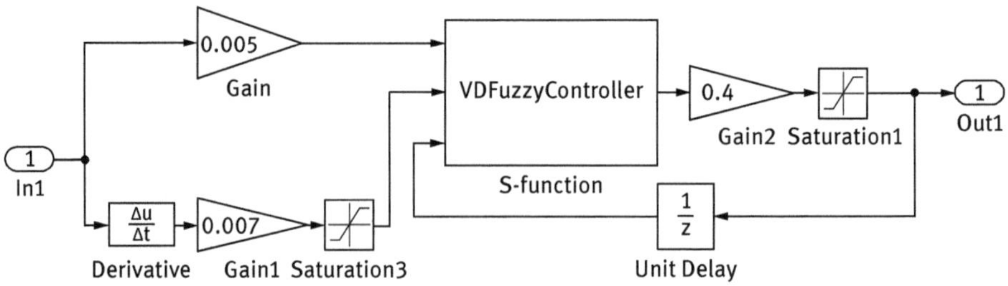

Figure 4.29: Simulink diagram of real-time simplified VDFLC algorithm subsystem.

3) C-Sfunction for real-time simplified VDFuzzyController

```
/*C-Sfunction: VDFuzzyControl.c====================================
 * Abstract:
 * This C-Sfunction is used to simulate the real-time simplified variable
domain
 * fuzzy control algorithm.
 */
 #define S_FUNCTION_NAME VDFuzzyControl
 #define S_FUNCTION_LEVEL 2

/*
 * Need to include simstruc.h for the definition of the SimStruct and
 * its associated macro definitions.
 */
 #include "simstruc.h"
 #include "math.h"

 #define sampletime(S) ssGetSFcnParam(S, 0)

/*====================*
 * S-function methods *
 *====================*/

/*Function: mdlInitializeSizes ======================================
 * Abstract:
 *  The sizes information is used by Simulink to determine the S-function
 *  block's characteristics (number of inputs, outputs, states, etc.).
 */
 static int_T count;
```

```
static void mdlInitializeSizes(SimStruct *S)
{
 ssSetNumSFcnParams(S, 1);/* Number of expected parameters */
 if (ssGetNumSFcnParams(S) != ssGetSFcnParamsCount(S)) {
     /* Return if number of expected != number of actual parameters */
     return;
}

ssSetNumContStates(S, 0);
ssSetNumDiscStates(S, 0);

if (!ssSetNumInputPorts(S, 3)) return;  //Specify the number of input ports
that a block has.
 ssSetInputPortWidth(S, 0, 1);    //Specify the width of an input port.
 ssSetInputPortWidth(S, 1, 1);    //Specify the width of an input port.
 ssSetInputPortWidth(S, 2, 1);    //Specify the width of an input port.
 ssSetInputPortRequiredContiguous(S, 0, true);/*direct input signal access*/
 ssSetInputPortRequiredContiguous(S, 1, true);/*direct input signal access*/
 ssSetInputPortRequiredContiguous(S, 2, true);/*direct input signal access*/
 ssSetInputPortDirectFeedThrough(S, 0, 1);//Specify the direct feedthrough
status of a block's ports.
 ssSetInputPortDirectFeedThrough(S, 1, 1);//Specify the direct feedthrough
status of a block's ports.
 ssSetInputPortDirectFeedThrough(S, 2, 1);//Specify the direct feedthrough
status of a block's ports.

if (!ssSetNumOutputPorts(S, 1)) return;
ssSetOutputPortWidth(S, 0, 1);

ssSetNumSampleTimes(S, 1); //Specify the number of sample times that an
S-Function block has.
 ssSetNumRWork(S, 0);
 ssSetNumIWork(S, 0);
 ssSetNumPWork(S, 0);
 ssSetNumModes(S, 0);
 ssSetNumNonsampledZCs(S, 0);
 ssSetOptions(S, 0);
}
/*Function: mdlInitializeSampleTimes ================================
 * Abstract:
 *  This function is used to specify the sample time(s) for your
 *  S-function. You must register the same number of sample times as
 *  specified in ssSetNumSampleTimes.
```

```
*/
static void mdlInitializeSampleTimes(SimStruct *S)
{
   const real_T  *stime=mxGetPr(sampletime(S));
   ssSetSampleTime(S, 0, stime[0]);
   ssSetOffsetTime(S, 0, 0.0);
}

#define MDL_INITIALIZE_CONDITIONS/* Change to #undef to remove function */
#if defined(MDL_INITIALIZE_CONDITIONS)
/*Function: mdlInitializeConditions ===================================
 * Abstract:
 *  In this function, you should initialize the continuous and discrete
 *  states for your S-function block. The initial states are placed
 *  in the state vector, ssGetContStates(S) or ssGetRealDiscStates(S).
 *  You can also perform any other initialization activities that your
 *  S-function may require. Note, this routine will be called at the
 *  start of simulation and if it is present in an enabled subsystem
 *  configured to reset states, it will be call when the enabled subsystem
 *  restarts execution to reset the states.
 */
static void mdlInitializeConditions(SimStruct *S)
{
}
#endif/* MDL_INITIALIZE_CONDITIONS */

#define MDL_START/* Change to #undef to remove function */
#if defined(MDL_START)
/*Function: mdlStart ==============================================
 * Abstract:
 *  This function is called once at start of model execution. If you
 *  have states that should be initialized once, this is the place
 *  to do it.
 */
static void mdlStart(SimStruct *S)
{
  count=0;
}
#endif/* MDL_START */

/*Function: mdlOutputs ============================================
 * Abstract:
```

```
* In this function, you compute the outputs of your S-function
* block. Generally outputs are placed in the output vector, ssGetY(S).
*/
static void mdlOutputs(SimStruct *S, int_T tid)
{
 const real_T *e = ssGetInputPortRealSignal(S,0);
 const real_T *de = ssGetInputPortRealSignal(S,1);
 const real_T *lu = ssGetInputPortRealSignal(S,2);
 real_T *u = (real_T *)ssGetOutputPortSignal(S,0);
 const real_T *stime=mxGetPr(sampletime(S));
 static real_T emin=-2,emax=2,demin=-2,demax=2;
 real_T SetC[7]={-3,-2,-1,0,1,2,3};
 real_T SetA[15]={-2, -1.33, -1.33, -0.66, -0.66, -0.66, -0, 0, 0, 0.66,
0.66, 0.66, 1.33, 1.33, 2};
 real_T SetB[15]={-2, -1.33, -1.33, -0.66, -0.66, -0.66, -0, 0, 0, 0.66,
0.66, 0.66, 1.33, 1.33, 2};
 real_T A[5]={0,0,0,0,0};
 real_T B[7]={0,0,0,0,0,0,0};
 int_T i,j,k,pd1,pd2,pd3,pd4;
 real_T a0,a1,b0,b1,T0,T1,T2,T3,jz1[7],jz2[7];
 real_T jz3[7],jz4[7],Set_A[15],Set_B[15],range[7];
 int_T ii;
 real_T result[7],tmpmax1,tmpmax2;
 real_T sum1=0,sum2=0,tmpe,tmpde;

//////The Third Table: du//////
 static real_T C[7][7]={{1,0.7,0.3,0,0,0,0},
               {0.7,1,0.7,0.3,0,0,0},
               {0.3,0.7,1.0,0.7,0.3,0,0},
               {0,0.3,0.7,1,0.7,0.3,0},
               {0,0,0.3,0.7,1,0.7,0.3},
               {0,0,0,0.3,0.7,1,0.7},
               {0,0,0,0,0.3,0.7,1}};
//////The End of The Third Table//////

//////The Forth Table: Control Rule//////
 static int_T ConRule[5][5]={{3,3,0,3,3},
               {3,3,1,3,3},
               {0,1,3,5,6},
               {3,3,5,3,3},
               {3,3,6,3,3}};
//////The End of The Forth Table//////
```

```
tmpe=e[0];
tmpde=de[0];
if (count!=0)
{
 for (i=0;i<15;i++)
  Set_A[i]=pow(fabs(tmpe)/2,0.1)*SetA[i];
 for (j=0;j<15;j++)
  Set_B[j]=pow(fabs(tmpde)/2,0.1)*SetB[j];
 for (k=0;k<7;k++)
  range[k]=pow(pow(fabs(tmpe)/2,0.1)*pow(fabs(tmpde)/2,0.1),0.01)
  *SetC[k];
}
else
{
 for (i=0;i<15;i++)
  Set_A[i]=SetA[i];
 for (j=0;j<15;j++)
  Set_B[j]=SetB[j];
 for (k=0;k<7;k++)
  range[k]=SetC[k];
}
count=count+1;

//////The First Table: e//////
if (tmpe>=Set_A[0]&&tmpe<Set_A[1])
A[0]=1;
else if (tmpe>=Set_A[1]&&tmpe≤Set_A[3])
A[0]=-1/(Set_A[3]-Set_A[1])*(tmpe-Set_A[1])+1;
else;
if (tmpe>=Set_A[2]&&tmpe<Set_A[4])
A[1]=1/(Set_A[4]-Set_A[2])*(tmpe-Set_A[4])+1;
else if (tmpe>=Set_A[4]&&tmpe<Set_A[6])
A[1]=-1/(Set_A[6]-Set_A[4])*(tmpe-Set_A[4])+1;
else;
if (tmpe>=Set_A[5]&&tmpe<Set_A[7])
A[2]=1/(Set_A[7]-Set_A[5])*(tmpe-Set_A[7])+1;
else if (tmpe>=Set_A[7]&&tmpe<Set_A[9])
A[2]=-1/(Set_A[9]-Set_A[7])*(tmpe-Set_A[7])+1;
else;
if (tmpe>=Set_A[8]&&tmpe<Set_A[10])
A[3]=1/(Set_A[10]-Set_A[8])*(tmpe-Set_A[10])+1;
else if (tmpe>=Set_A[10]&&tmpe<Set_A[12])
```

```
 A[3]=-1/(Set_A[12]-Set_A[10])*(tmpe-Set_A[10])+1;
 else;
 if (tmpe>=Set_A[11]&&tmpe<Set_A[13])
 A[4]=1/(Set_A[13]-Set_A[11])*(tmpe-Set_A[13])+1;
 else if (tmpe>=Set_A[13]&&tmpe<Set_A[14])
 A[4]=1;
 else;
//////The End of The First Table//////

//////The Second Table: de//////
 if (tmpde>=Set_B[0]&&tmpde<Set_B[1])
 B[0]=1;
 else if (tmpde>=Set_B[1]&&tmpde≤Set_B[3])
 B[0]=-1/(Set_B[3]-Set_B[1])*(tmpde-Set_B[1])+1;
 else;
 if (tmpde>=Set_B[2]&&tmpde<Set_B[4])
 B[1]=1/(Set_B[4]-Set_B[2])*(tmpde-Set_B[4])+1;
 else if (tmpde>=Set_B[4]&&tmpde<Set_B[6])
 B[1]=-1/(Set_B[6]-Set_B[4])*(tmpde-Set_B[4])+1;
 else;
 if (tmpde>=Set_B[5]&&tmpde<Set_B[7])
 B[2]=1/(Set_B[7]-Set_B[5])*(tmpde-Set_B[7])+1;
 else if (tmpde>=Set_B[7]&&tmpde<Set_B[9])
 B[2]=-1/(Set_B[9]-Set_B[7])*(tmpde-Set_B[7])+1;
 else;
 if (tmpde>=Set_B[8]&&tmpde<Set_B[10])
 B[3]=1/(Set_B[10]-Set_B[8])*(tmpde-Set_B[10])+1;
 else if (tmpde>=Set_B[10]&&tmpde<Set_B[12])
 B[3]=-1/(Set_B[12]-Set_B[10])*(tmpde-Set_B[10])+1;
 else;
 if (tmpde>=Set_B[11]&&tmpde<Set_B[13])
 B[4]=1/(Set_B[13]-Set_B[11])*(tmpde-Set_B[13])+1;
 else if (tmpde>=Set_B[13]&&tmpde<Set_B[14])
 B[4]=1;
 else;
//////The End of The Second Table//////

 for (i=0;i<4;i++)//Find the nonzero sets
 {
 if (A[i]==0)
   continue;
 else
```

```
{
  a0=A[i];
  a1=A[i+1];
  break;
 };
}
for (j=0;j<4;j++)
  {if (B[j]==0)
   continue;
  else
  {
    b0=B[j];
    b1=B[j+1];
    break;
  };
}
if (i>3)i=3;
if (j>3)j=3;
pd1=ConRule[j][i];//look up rules, and get member functions
pd2=ConRule[j+1][i+1];
pd3=ConRule[j+1][i];
pd4=ConRule[j][i+1];
if(a0>=b0) T0=b0;
else
 T0=a0;
if(a1>=b1) T1=b1;
else
 T1=a1;
if(a0>=b1) T2=b1;
else
 T2=a0;
if(a1>=b0) T3=b0;
else
 T3=a1;

switch (pd1)
{
case 0:
 for(ii=0;ii<7;ii++)
 jz1[ii]=C[0][ii];
 break;
```

```
case 1:
 for(ii=0;ii<7;ii++)
 jz1[ii]=C[1][ii];
 break;
case 2:
 for(ii=0;ii<7;ii++)
 jz1[ii]=C[2][ii];
 break;
case 3:
 for(ii=0;ii<7;ii++)
 jz1[ii]=C[3][ii];
 break;
case 4:
 for(ii=0;ii<7;ii++)
 jz1[ii]=C[4][ii];
 break;
case 5:
 for(ii=0;ii<7;ii++)
 jz1[ii]=C[5][ii];
 break;
case 6:
 for(ii=0;ii<7;ii++)
 jz1[ii]=C[6][ii];
 break;
 default:
 break;
}
switch (pd2)
{
case 0:
 for(ii=0;ii<7;ii++)
 jz2[ii]=C[0][ii];
 break;
case 1:
 for(ii=0;ii<7;ii++)
 jz2[ii]=C[1][ii];
 break;
case 2:
 for(ii=0;ii<7;ii++)
 jz2[ii]=C[2][ii];
 break;
```

```
case 3:
 for(ii=0;ii<7;ii++)
 jz2[ii]=C[3][ii];
 break;
case 4:
 for(ii=0;ii<7;ii++)
 jz2[ii]=C[4][ii];
 break;
case 5:
 for(ii=0;ii<7;ii++)
 jz2[ii]=C[5][ii];
 break;
case 6:
 for(ii=0;ii<7;ii++)
 jz2[ii]=C[6][ii];
 break;
default:
 break;
}
switch (pd3)
{
case 0:
 for(ii=0;ii<7;ii++)
 jz3[ii]=C[0][ii];
 break;
case 1:
 for(ii=0;ii<7;ii++)
 jz3[ii]=C[1][ii];
 break;
case 2:
 for(ii=0;ii<7;ii++)
 jz3[ii]=C[2][ii];
 break;
case 3:
 for(ii=0;ii<7;ii++)
 jz3[ii]=C[3][ii];
 break;
case 4:
 for(ii=0;ii<7;ii++)
 jz3[ii]=C[4][ii];
 break;
```

```
case 5:
 for(ii=0;ii<7;ii++)
 jz3[ii]=C[5][ii];
 break;
case 6:
 for(ii=0;ii<7;ii++)
 jz3[ii]=C[6][ii];
 break;
 default:
 break;
}
switch (pd4)
{
case 0:
 for(ii=0;ii<7;ii++)
 jz4[ii]=C[0][ii];
 break;
case 1:
 for(ii=0;ii<7;ii++)
 jz4[ii]=C[1][ii];
 break;
case 2:
 for(ii=0;ii<7;ii++)
 jz4[ii]=C[2][ii];
 break;
case 3:
 for(ii=0;ii<7;ii++)
 jz4[ii]=C[3][ii];
 break;
case 4:
 for(ii=0;ii<7;ii++)
 jz4[ii]=C[4][ii];
 break;
case 5:
 for(ii=0;ii<7;ii++)
 jz4[ii]=C[5][ii];
 break;
case 6:
 for(ii=0;ii<7;ii++)
 jz4[ii]=C[6][ii];
 break;
```

```
  default:
  break;
 }
 for(ii=0;ii<7;ii++)
   if(jz1[ii]>=T0)
     jz1[ii]=T0;
 for(ii=0;ii<7;ii++)
   if(jz2[ii]>=T1)
     jz2[ii]=T1;
 for(ii=0;ii<7;ii++)
   if(jz3[ii]>=T2)
     jz3[ii]=T2;
 for(ii=0;ii<7;ii++)
   if(jz4[ii]>=T3)
     jz4[ii]=T3;
 for (ii=0;ii<7;ii++)
 {
   if(jz1[ii]>=jz2[ii]) tmpmax1=jz1[ii];
   else
     tmpmax1=jz2[ii];
   if(tmpmax1>=jz3[ii]) tmpmax2=tmpmax1;
   else
     tmpmax2=jz3[ii];
   if(tmpmax2>=jz4[ii]) result[ii]=tmpmax2;
   else
     result[ii]=jz4[ii];
 }
 for (ii=0;ii<7;ii++)
  {
   sum1=sum1+result[ii]*range[ii];
   sum2=sum2+result[ii];
  }
   u[0]=lu[0]/0.4+sum1/sum2;
 }

 #define MDL_UPDATE/* Change to #undef to remove function */
 #if defined(MDL_UPDATE)
/*Function: mdlUpdate ==========================================
 * Abstract:
 *  This function is called once for every major integration time step.
 *  Discrete states are typically updated here, but this function is useful
 *  for performing any tasks that should only take place once per
```

```
 * integration step.
 */
 static void mdlUpdate(SimStruct *S, int_T tid)
 {
 }
 #endif/* MDL_UPDATE */

 #define MDL_DERIVATIVES/* Change to #undef to remove function */
 #if defined(MDL_DERIVATIVES)
/*Function: mdlDerivatives ==========================================
 * Abstract:
 *  In this function, you compute the S-function block's derivatives.
 *  The derivatives are placed in the derivative vector, ssGetdX(S).
 */
 static void mdlDerivatives(SimStruct *S)
 {
 }
 #endif/* MDL_DERIVATIVES */

/*Function: mdlTerminate ============================================
 * Abstract:
 *  In this function, you should perform any actions that are necessary
 *  at the termination of a simulation. For example, if memory was
 *  allocated in mdlStart, this is the place to free it.
 */
 static void mdlTerminate(SimStruct *S)
 {
 }

/*=======================================================*
 * See sfuntmpl_doc.c for the optional S-function methods *
 *=======================================================*/

/*==============================*
 * Required S-function trailer *
 *==============================*/

 #ifdef MATLAB_MEX_FILE /* Is this file being compiled as a MEX-file? */
 #include “simulink.c” /* MEX-file interface mechanism */
 #else
 #include “cg_sfun.h ”/* Code generation registration function */
 #endif
```

4) Program running results

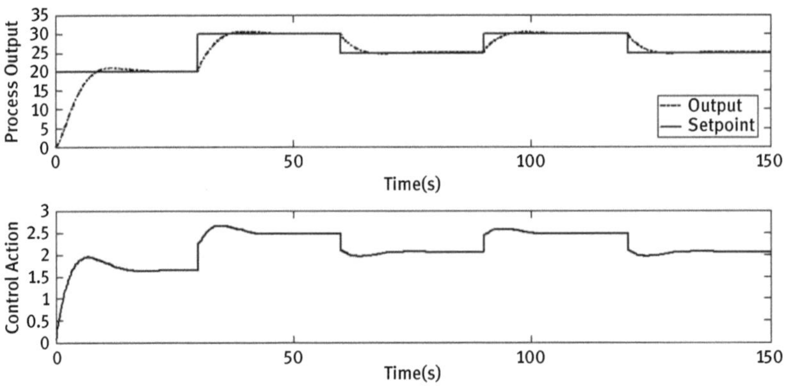

Figure 4.30: Simulative result of real-time simplified VDFLC algorithm.

References

[1] Tong, S. W., G. P. Liu, X. G. Wang and M. Tan. 2009. Real-time implementation of adaptive state feedback predictive control of PEM fuel cell flow systems using the singular pencil model method. IEEE Transactions on Control Systems Technology.17: 697–706.

[2] Tong, S. W. and G. P. Liu. 2008. Real-time simplified variable domain fuzzy control of PEM fuel cell flow systems. European Journal of Control. 14: 223–233.

[3] Salut,G., J. Aquilar-Martin and S. Lefebvre.1979. Canonical input-output representation of linear multivariable stochastic systems and joint optimal parameter and state estimation. Stochastica. 111: 17–38.

[4] Salut,G., J. Aquilar-Martin and S. Lefebvre.1980. New results on optimal joint parameter and state estimation of linear stochastic systems. ASME Transaction Jounal of Dynamic System Measurement and Control. 102: 28–34.

[5] Wang, X. G. 1995. Self-tuning control with singular pencil models. In: 1995 IEEE International Conference on System, Man and Cybernetics. Vancouver, Canada. 22–25 October 1995. IEEE. Piscataway, USA.

[6] Wang, X. G. and W. Liu. 2000. Singular pencil model-based predictive control strategy. In: Proceedings of the 2000 IEEE International Conference on Control Applications. Alaska, USA. 25–27 September 2000. IEEE. Piscataway, USA.

[7] Wang, X. G. and W. Liu. 2001. A singular pencil model fault diagnosis strategy for an industrial process. In: IEEE Systems Readiness Technology Conference, 2001 IEEE AUTOTESTCON Proceedings. Valley Forge, USA. 20–23 August 2001. IEEE. Piscataway, USA.

[8] Aplevich, J. D. 1981. Time-Domain input-output representations of linear systems. Automatica. 17: 509–521.

[9] Aplevich, J. D. 1982. Application of Singular Pencil Representations to Control System Design. IEEE Control System Magazine. 2: 46–46.

[10] Aplevich, J. D. 1985. Minimal representations of implicit linear systems. Automatica. 21: 259–269.

[11] Aplevich, J. D. 1991. Implicit Linear Systems. Lecture Notes in Control and Information Sciences. Springer-Verlag. Heidelberg, Germany.
[12] Saridis, G. N. 1974. Stochastic approximation methods for identification and control-a survey. IEEE Transactions on Automatic Control. 19: 798–809.
[13] Chen, Y. C., J. D. Aplevich and W. J. Wilson. 1986. Simultaneous estimation of state and parameters for multivariable linear systems with singular pencil models. IEE Proceedings-Control Theory and Applications. 133: 65–72.
[14] Camacho, E. F. and C. Bordons. 1998. Model Predictive Control. Springer-Verlag. Heidelberg, Germany.
[15] Richalet, J., S. A. Ata-Doss, C. Arber, H. B. Kuntze, A. Jacubasch and W. Schill. 1987. Predictive functional control: application to fast and accurate robots. In: Proceedings of the 10th IFAC World Congress. Munich, Germany. 27–31 July 1987.
[16] Hu, P. H., X. Zuo and P. Yuan. 2000. Properties of state feedback model predictive control systems. In: Proceedings of the 3rd World Congress on Intelligent Control and Automation. Hefei, China. 26 June–2 July 2000. IEEE. Piscataway, USA.
[17] Gawthrop, P. J., H. Demircioglu and I. I. Siller-Alcala. 1998. Multivariable continuous-time generalized predictive control: A state-space approach to linear and nonlinear systems. IEE Proceedings-Control Theory and Applications. 145: 241–250.
[18] Soeterboek, A. R. M., H. B. Verbruggen, d. B. P. P. J. Van and H. Butler. 1990. On the unification of predictive control algorithms. In: Proceedings of the 29th IEEE Conference on Control and Decision. Wawaii, USA. 5–7 December 1990. IEEE. Piscataway, USA.
[19] Clarke, D. W., C. Mohtadi and P. S. Tuffs. 1987. Generalized predictive control-part I the basic algorithm, and part II extension and interpretations. Automatica. 23: 137–160.
[20] Cutler, H. and B. L. Ramaker. 1980. Dynamic matrix control-a computer control algorithm. In: Proceedings of the Joint Automatic control conference. San Francisco, USA. January 1980. IEEE. Piscataway, USA.
[21] Cong, S. 2001. Neural network, fuzzy system and its application in motion control. China University of Science and Technology Press. Anhui Hefei.
[22] Huang, Y., Y. Q. Yu and T. Zeng. 2005. A new real-time self-adaptive rule modification algorithm based on error convergence in fuzzy control. 2005 IEEE Conference on Industrial Technology. Hong Kong, China. 14–17 December 2005. IEEE. Piscataway, USA.
[23] Liu, Z. L., C. Y. Su and J. Svoboda. 2004. Control of wing rock phenomenon with a variable universe fuzzy controller. In: Proceeding of the 2004 American Control Conference. Massachusetts, USA. 30 June–2 July 2004. IEEE. Piscataway, USA.
[24] Li, H. X. 1995. To see the success of fuzzy logic from the mathematical essence of fuzzy controls. Fuzzy System and Mathematics. 9: 1–14.
[25] Li, H. X. 1997. The mathematical nature of fuzzy control and the design of a class of high-precision fuzzy controller. Control Theory and Application. 14: 868–872.
[26] Li, H. X. 1997. The mathematical essence of fuzzy controls and fine fuzzy controllers. Advances in Machine Intelligence and Soft-Computing. Edited by P. P. Wang. Bookwright Press. Durham.
[27] Li, H. X. 1999. Adaptive fuzzy controllers based on variable universe. Science in China (Series E). 42: 10–20.
[28] Li, H. X. and C. P. Chen. 2000. The equivalence between fuzzy logic systems and feedforward neural networks. IEEE Transactions on Neural Networks. 11: 356–365.
[29] Li, H. X., Z. H. Miao and J. Y. Wang. 2002. Variable universe adaptive fuzzy control on the quadruple inverted pendulum. Science in China (Series E). 45: 213–224.
[30] Li, H. X., Z. H. Miao and E. S. Lee. 2002. Variable universe stable adaptive fuzzy control of a nonlinear system. Computers and Mathematics with Applications. 44: 799–815.

[31] Oh, S. Y. and D. J. Park. 1995. Self-tuning fuzzy controller with variable universe of discourse. In: 1995 IEEE International Conference on Systems, Man and Cybernetics. Vancouver, Canada. 22–25 October 1995. IEEE. Piscataway, USA.
[32] Yang, X. Y., Q. X. You and H. X. Li. 2005. Automatic reversing control based on the variable universe theory. Journal of Beijing Normal University (Natural Science Edition). 41: 348–350.

5 Control of PEM fuel cell with load current feedback

The design of the control system will depend on different requirements according to the different hardware structure of the fuel cell. One of the main concerns for us in the Chinese Academy of Sciences is the cost in designing of 1 kW hydrogen-air proton exchange membrane (PEM) fuel cell device. The inlet of the anode or the cathode channel is equipped with a flowmeter as a means of automatic adjustment, and the outlet of the anode or the cathode is only equipped with a manual ball valve, lacking means of automatic adjustment, which forms an underactuated control system. The pressure and flow of the cathode and the anode need to be controlled simultaneously so as to provide power for the load and prevent premature aging of the stack. For the cathode channel, the control problem lies in the use of an operation variable to regulate the two controlled variables. On the other hand, because of the complex mechanism of the fuel cell, it is difficult to build an accurate mathematical model. This chapter applies the nonmodel control method proposed by the researcher of the Chinese Academy of Sciences, Jianqiang Yi [1–4], to achieve the control of the PEM fuel cell flow underactuated system considering the feedback of load current [5].

5.1 Control approach for underactuated systems based on fuzzy reasoning and random search optimization

5.1.1 Dynamically connected fuzzy inference model based on the single-input rule modules

Considering a n-input 1-output system, we can define a single-input rule module (SIRM) for each input by

$$\mathbf{SIRM-i}:\left\{R_i^j:\text{if } x_i=A_i^j \text{ then } \Delta u_i=C_i^j\right\}_{j=1}^{m_i} \tag{5.1}$$

where SIRM-i is the the ith input of the SIRM, R_i^j is the jth rule of SIRM-i, the ith input item x_i is an antecedent variable, and the consequent variable Δu_i is an intermediate variable corresponding to the output item Δu, A_i^j and C_i^j are the membership functions of x_i and Δu_i, respectively, i is the number of SIRM $i=1,2,\cdots,n$, j is the rule number of the SIRM-i, $j=1,2,\cdots,m$.

Using the simplified reasoning method, the inference results Δu_i^0 for the consequent variable Δu_i of the SIRM-i is:

$$\Delta u_i^0=\frac{\sum_{j=1}^{m_i}A_i^j(x_i)C_i^j}{\sum_{j=1}^{m_i}A_i^j(x_i)} \tag{5.2}$$

In general, each input item plays a different role in controlling performance. Some of the items contribute more to control performance, while others are relatively small.

https://doi.org/10.1515/9783110602159-005

Enhancing the effect of those items that have a greater contribution to control performance will help improve control performance while enhancing those items with the smaller contribution to the control performance will have little effect on the improvement of the control performance. This can be done by assigning the former a large weight to enhance their contributions while giving the latter a small weight to restrict their actions. This will help improve the overall control performance.

The concept of Importance Degree (ID) is introduced into the fixed model of the SIRM to characterize the effect of the input term on the control performance. However, in the entire control process, the ID is kept fixed, even for different stages of the first order system, such as the ascending and stabilizing stage. Although this will make the structure of the fuzzy controller simple, it is difficult to further improve the control performance. Taking the first order plus pure lag process as an example, and using the output error and error changes as input items and using two important indicators of the rise time and overshoot to evaluate the system control performance, we can see that the two indicators are usually contradictory. If you want to shorten the rise time, it is necessary to increase the ID of the output error. But because the ID is fixed, larger values will result in larger overshoots. If we use a larger ID at the beginning of the control, and gradually decrease the ID as the output approaching the reference, the shorter rise time and the smaller overshoot could be achieved at the same time. This means that the ID should be dynamically adjusted according to the different conditions for improving the control performance. Yi et al. proposed a SIRM-based dynamically connected fuzzy inference model [1–4]. One of the significant features of their model is the introducing of the concept Dynamic Importance Degree (DID). They defined a DID ω_i^D for each input item x_i:

$$\omega_i^D = \omega_i + B_i \Delta\omega_i^0 \tag{5.3}$$

where ω_i represents the base value, ensuring that each input item can function throughout the control process. The range of dynamic variable $\Delta\omega_i$ is between 0 and 1, reflecting different control situations and requiring real-time dynamic adjustment. Dynamic value is represented by the production of bandwidth B_i and the inference result $\Delta\omega_i^0$ of the dynamic variable $\Delta\omega_i$. Bandwidth B_i can restrict the upper limit of the change of the dynamic value. Then, the dynamic value can be adjusted according to different situations. The corresponding change range of the DI is $[\omega_i, \omega_i + B_i]$. The base value ω_i and the bandwidth B_i are adjustable parameters, which can be obtained by the random search optimization method such that a cost index J reaches global minimum. J can be an indicator of the cumulative error of the system.

In this way, the output item Δu can be described by each dynamic importance ω_i^D and the inference result of each SIRM Δu_i^0 as:

$$\Delta u = \sum_{i=1}^{n} \omega_i^D \Delta u_i^0 \tag{5.4}$$

Because the output is a linear combination of DID, those inputs with greater DID add greater contribution to the system, while those inputs with smaller DID have less importance to the system. Furthermore, the relationship between the input items defined by the importance degree is also linearly mapped to the output of the model. Therefore, we can consciously adjust the importance degree to achieve a control object. This process is a bit like tuning PID parameters. But the dynamic importance is usually nonlinear function of the corresponding input item.

If the bandwidths of all importance degrees are set to 0, the SIRM-based dynamically connected fuzzy inference model will essentially become a fixed model based on SIRM. Thus, the fixed model based on SIRM can be regarded as a special case of SIRM-based dynamically connected fuzzy inference model.

5.1.2 Random search optimization

1. Basic algorithm

In a variety of optimization algorithms, the random search optimization algorithm has been widely used for its simple calculation, adapting to complex objective functions and global optimization. This method is very suitable for real-time application because it is a variable step method and does not need to calculate the gradient [6].

Let $J(x)$ be a nonnegative objective function; the goal of optimization is to make it the smallest. Ω is the search area, and the search direction is determined by a random vector $\xi \sim N\left(\mu^{(k)}, \sum \mu^{(k)}\right)$.

– Step 1:
Initialize $x^{(0)} = 0, \mu^{(0)} = 0, k = 0,$ and M which is the maximum number of iterations.
– Step 2:
Generate a Gaussian distribution vector $\xi^{(k)}$; if $x^{(k)} \pm \xi^{(k)} \in \Omega$, execute step 3, otherwise execute step 4.
– Step 3:

(1) If $J(x^{(k)} + \xi^{(k)}) < J\left(x^{(k)}\right)$, then

$$\begin{cases} x^{(k+1)} = x^{(k)} + \xi^{(k)} \\ \mu^{(k+1)} = 0.4\xi^{(k)} + 0.2\mu^{(k)} \end{cases}$$

(2) If $J(x^{(k)} + \xi^{(k)}) \geq J\left(x^{(k)}\right)$, and $J(x^{(k)} - \xi^{(k)}) < J\left(x^{(k)}\right)$, then

$$\begin{cases} x^{(k+1)} = x^{(k)} - \xi^{(k)} \\ \mu^{(k+1)} = \mu^{(k)} - 0.4\xi^{(k)} \end{cases}$$

else:

$$\begin{cases} x^{(k+1)} = x^{(k)} \\ \mu^{(k+1)} = 0.5\mu^{(k)} \end{cases}$$

– Step 4:
If $k = M$, the program is stopped; if $k < M$, let $k = k + 1$, and return to step 2.

2. Computer implementation

i. Off-line algorithm
Figure 5.1 is a block diagram of random search optimization off-line algorithm.

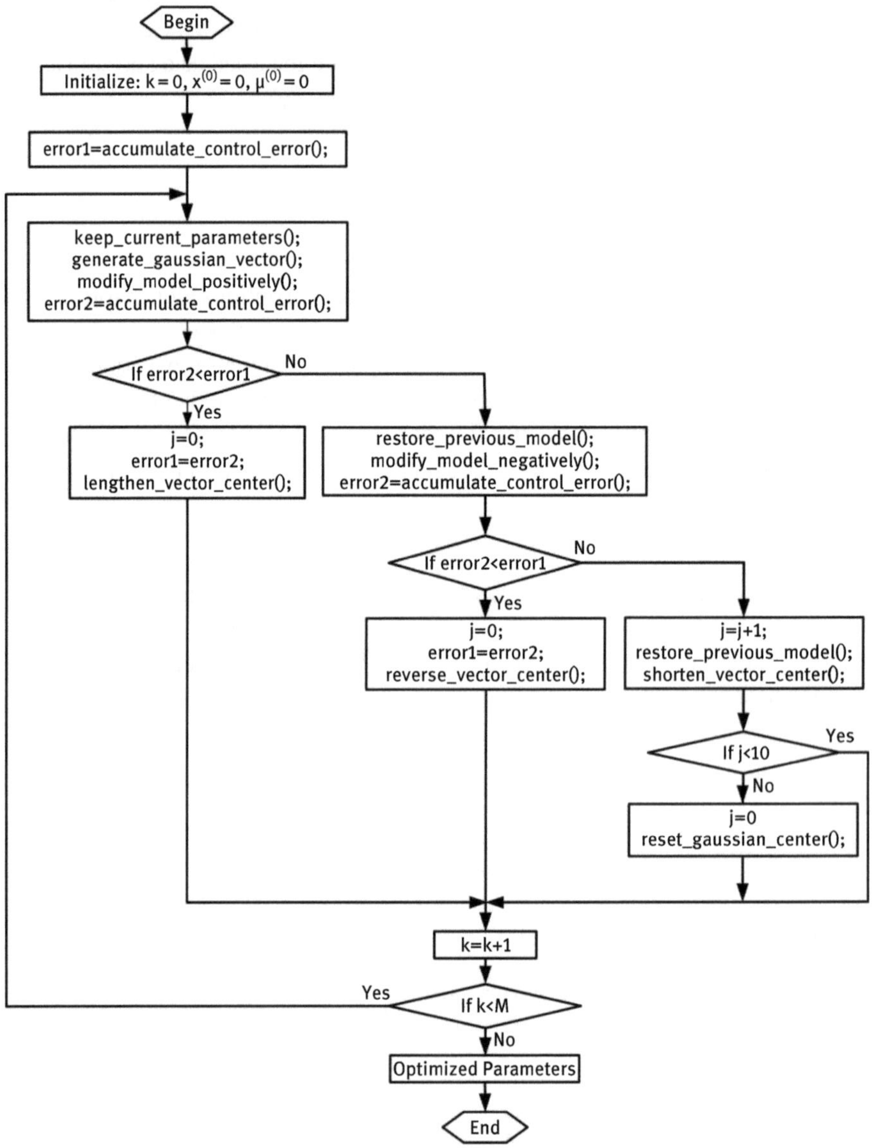

Figure 5.1: Structure diagram of the random search optimization off-line algorithm.

where,

Function accumulate_control_error( ) is a performance indicator function, used to calculate the cumulative error;
Function generate_gaussian_vector( ) is used to generate the Gaussian distribution matrix;
Function lengthen_vector_center( ) is used to adjust the Gaussian center in the positive direction;
Function reverse_vector_center( ) is used to adjust the Gaussian center in the negative direction;
Function shorten_vector_center( ) is used to narrow the center of the Gaussian in half;
Function reset_gaussian_Vector( ) is used to prevent getting into the local optimum and reset the Gaussian center;
Function modify_model_positively( ) is used to modify the dynamic importance degree in the positive direction;
Function modify_model_negatively( ) is used to modify the dynamic importance degree in the negative direction;
Function keep_current_parameters( ) is used to save the current parameters; and
Function restore_previous_model( ) is used to restore the last parameters.
Error2 is the calculated performance indicator at the current time. Error1 is the calculated performance indicator at the previous time. The direction of Gaussian search can be determined by comparing these two values.

ii. Online algorithm

Online optimization algorithm is based on the C-S function. Since the performance indicator can only be obtained in the next cycle after updating the optimized parameters, the Select Case structure as C language is utilized by judging CaseNo/NoStep to determine next performance. Three or four cycles' implementation is equivalent to one cycle implementation of off-line optimization algorithm. In this book, two versions of C-S functions for the online optimization algorithm are developed, where the function box_Muller(lxrv, lmiu) is used to generate the Gaussian distribution matrix and the function evaluate(xi, 0) is used to calculate the cost function of the optimization algorithm. The block diagrams are shown in Figures 5.2 and 5.3.

iii. Obtaining Gaussian distribution in C language

Since the stochastic search optimization algorithm uses the random variables of the Gaussian distribution and no random number function in the standard C language to directly produce Gaussian distribution, the Gaussian distribution random number should be obtained by a series of operations using the C language random number.

Box and Muller, in 1958, gave an algorithm for generating a Gaussian distribution random variable using uniform distribution random variables [7]. Let U_1 and

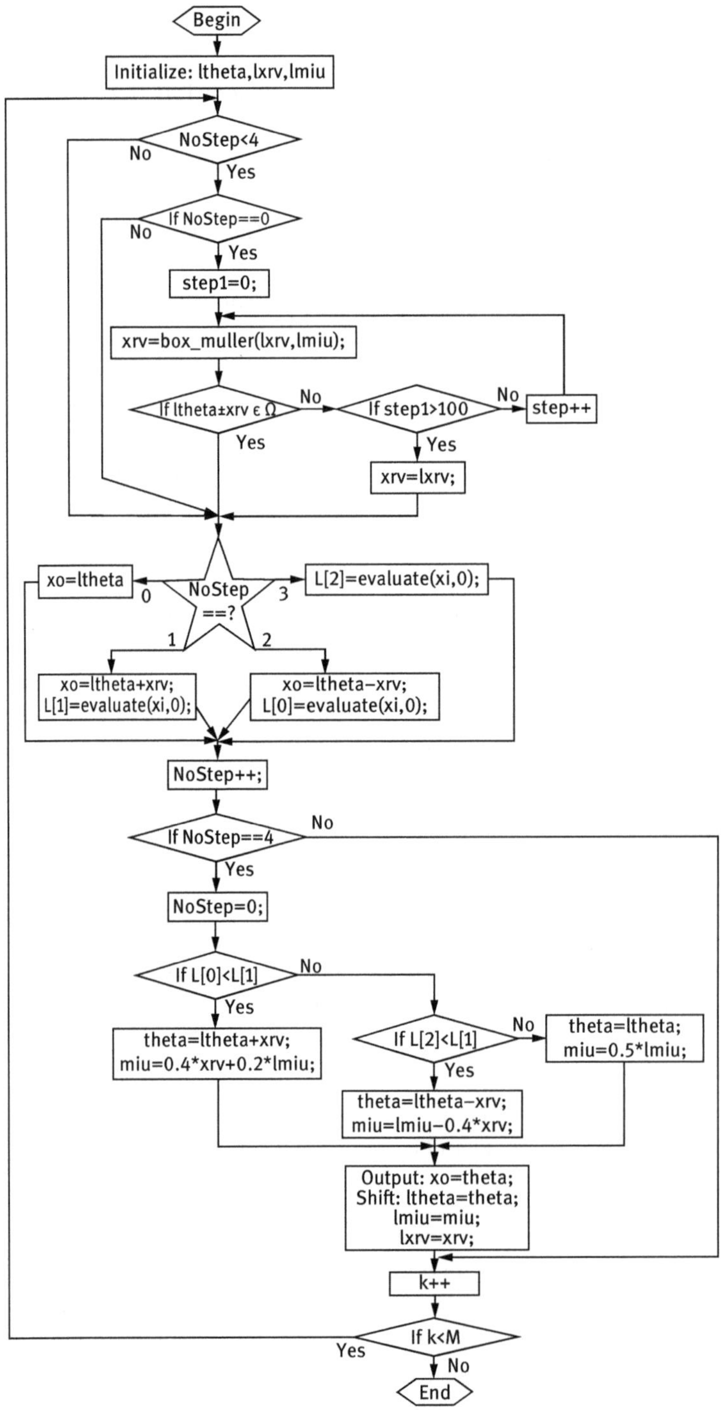

Figure 5.2: Structure diagram of the random search optimization online algorithm (Version 1).

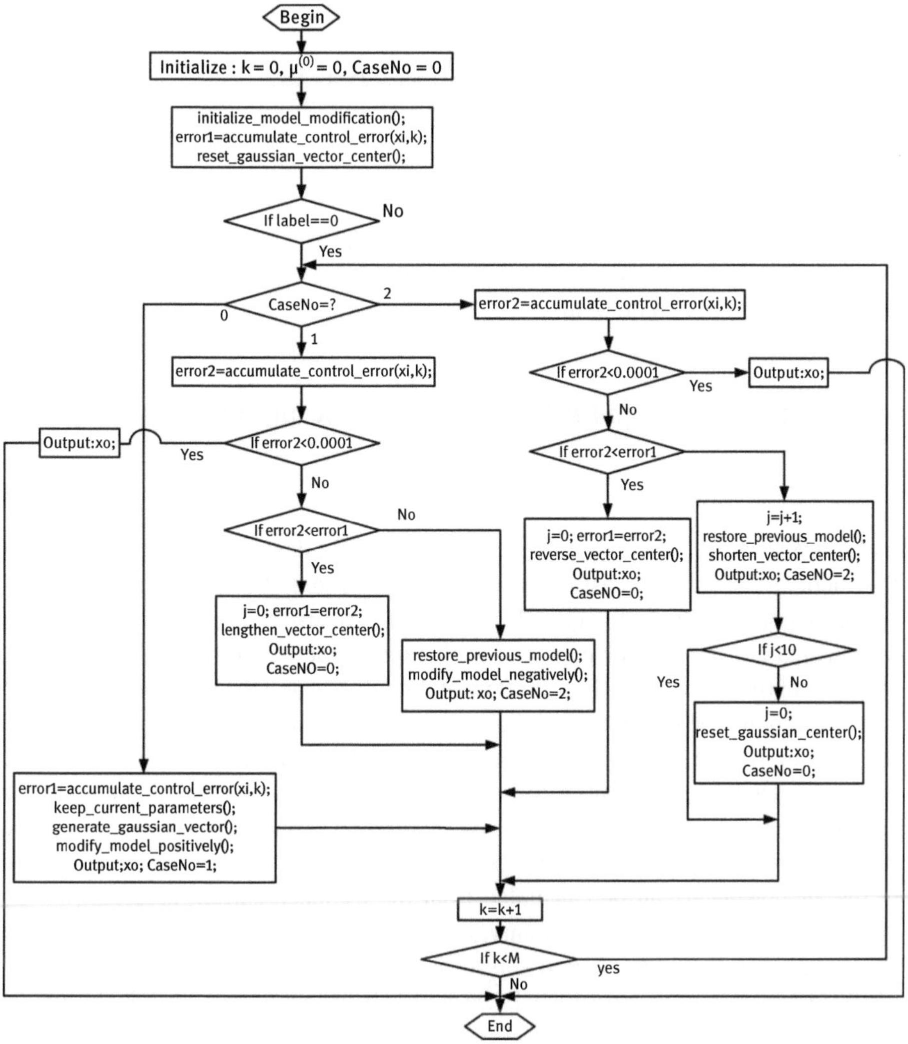

Figure 5.3: Structure diagram of the random search optimization online algorithm (Version 2).

U_2 be the independent uniform distribution random variables in section (0, 1), U_1 and U_2 are available through the rand( ) function provided by the standard C language:

$$U_1 = \text{rand}()/\text{RAND\_MAX}$$

$$U_2 = \text{rand}()/\text{RAND\_MAX} \tag{5.5}$$

Let

$$X_1 = \sqrt{-2 \times \lg(U_1)} \times \cos(2 \times \pi \times U_2)$$

$$X_2 = \sqrt{-2 \times \lg(U_1)} \times \sin(2 \times \pi \times U_2) \tag{5.6}$$

Then X_1 and X_2 are subject to $N(0,1)$ distribution and are independent of each other. In this way, we can obtain two independent $N(0,1)$ random numbers by using two independent $U(0,1)$ random numbers.

5.2 Controller design for PEM fuel cell flow underactuated system

5.2.1 Control strategy

In the 1kW PEM fuel cell flow system of the Chinese Academy of Sciences, the inlet of the anode or the cathode passage is equipped with a mass flowmeter, and the corresponding outlet is fitted with a hand valve and a pressure sensor. The hand valve can be adjusted but fixed in the course of each experiment. According to this hardware structure of the fuel cell, the inlet air flow of the cathode and the cathode pressure are treated as controlled variables, and the set voltage of the air flowmeter is considered as the operating variable; the anode pressure follows the cathode pressure changes. The control strategy actually forms a control problem of the underactuated system in the cathode flow system by controlling the two controlled variables (air flow and cathode pressure) with an operating variable (set voltage of the air flowmeter). The set voltage of the air flowmeter can be calculated from the feedback current according to the electrochemical principle. In summary, considering the fuel cell flow system with a load current feedback, the key to control is the control of the cathode flow underactuated system and the pressure balance control between the cathode and the anode.

5.2.2 Fuzzy controller design

In this section, we will realize the control of the fuel cell cathode flow underactuated system by using the SIRM-based dynamically connected fuzzy inference model proposed by Yi et al. [1–4]. First, we analyze the dynamic process of the cathode flow system: when the cathode pressure is lower than the given value, it should increase the set voltage of the air flowmeter to ensure the air increase into the cathode path, followed by an increase in the cathode pressure. Conversely, when the cathode pressure is higher than the given value, the set voltage of the air flowmeter should be reduced to guarantee the air reduction into the cathode, subsequently depressing the cathode pressure depressing.

The air flow can be seen as a disturbance to the cathode pressure control. The air flow control can be indirectly realized by consciously regulating the pressure in the opposite direction of the air flow control. Thus, the SIRMs for the air flow error and the variations of the error can be set as follows:

If the air flow rate is low, the set voltage of the air flowmeter will increase, which will lead to an increase in the cathode pressure. Conversely, if the air flow is high, the set voltage of the air flowmeter will decrease, which will lead to a decrease in the cathode pressure. Through the control of the cathode pressure, air flow control can be achieved indirectly.

From the above analysis, the fuzzy rules for each SIRM can be set as Table 5.1, where the consequent variable $\Delta u_i^0(k)$ is the intermediate variable of the controller output item $\Delta u_i(k)$. Using the simplified reasoning method, the membership function of the antecedent variable in each SIRM is defined as shown in Figure 5.4.

Table 5.1: SIRM for each input item in constant control.

Antecedent variable $x_i(k)$, $i = 1, 2, 3, 4$	Subsequent variable $\Delta u_i(k)$, $i = 1, 2, 3, 4$
NB	−1.0
ZO	0
PB	1.0

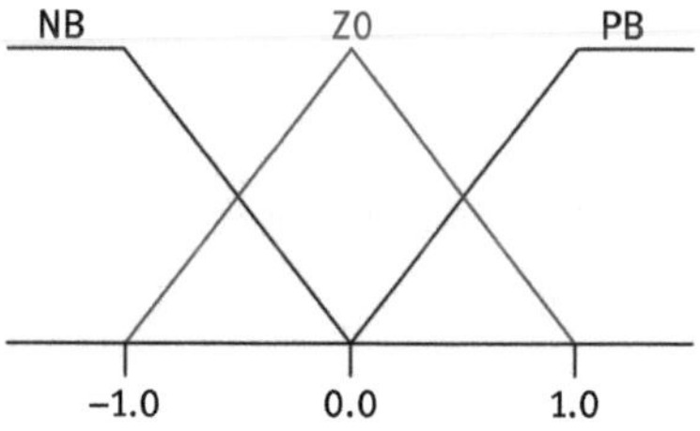

Figure 5.4: Membership function of the SIRM antecedent variable.

It can be seen from the proposed SIRM-based dynamically connected fuzzy inference model, a dynamic importance degree should be defined for each input item. If the air flow control level is higher than the cathode pressure control, the pressure control will fail. Therefore, the level of pressure control should be higher than air flow control, and air flow control should be carried out after the basic realization of the cathode pressure control. The level of control can be expressed in terms of dynamic importance degree; the greater the value of the e degree, the higher the level of the corresponding input item.

If the absolute value of the difference between the cathode pressure and the given value is large, it indicates that the cathode pressure is far from the desired position and that control should be applied as soon as possible. In this case, it should be set to a greater dynamic importance degree to enhance the control action. When the cathode pressure is quite close to the given value, it should reduce the dynamic importance degree. Therefore, the absolute value of the difference between the cathode pressure and the set value is chosen as the antecedent variable. The dynamic importance degrees of the cathode pressure error and the change of error are presented in Table 5.2. The membership functions of DS, DM, DB are shown in Figure 5.5.

Table 5.2: Control rules for cathode pressure dynamic variables in constant control.

Antecedent variable $\|x_1(k)\|$	**Subsequent variable** $\Delta\omega_i(k), i = 1, 2$
DS	0
DM	0.5
DB	1

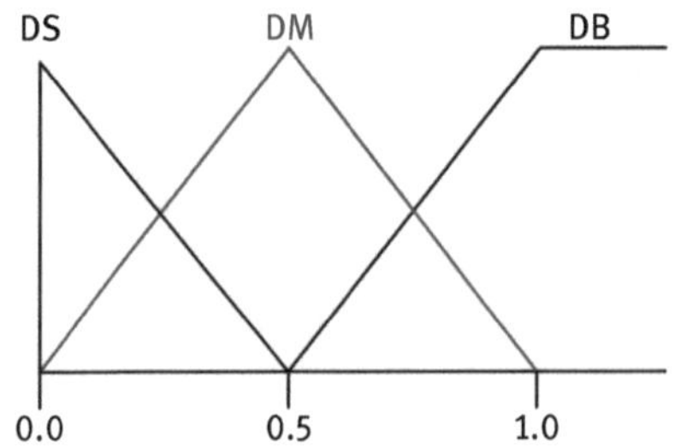

Figure 5.5: Membership function of the dynamic importance degree.

As mentioned earlier, in order to simultaneously control the cathode pressure and air flow, the control level of the cathode pressure must be higher than that of the air flow control. First to ensure the pressure control, the flow control can be started after the pressure control is basically completed. This fact means that a bigger dynamic important degree should be set to the pressure control than the flow control. The dynamic value should be determined by the absolute deviation of the air flow. When the pressure deviation is large, the pressure control should immediately function; in this case, the dynamic important degrees of the flow deviation and the variation of deviation should be reduced and the control should be mainly concentrated in the pressure control. When the pressure control is basically realized, the DIDs of flow

Table 5.3: Control rules for dynamic control of air flow in constant control.

Antecedent variable $\|x_3(k)\|$	Subsequent variable $\Delta\omega_i(k)$, $i = 3, 4$
DS	1
DM	0.5
DB	0

deviation and the variation of deviation should be appropriately increased to ensure the beginning of the air flow control such that the cathode pressure control is basically realized. Thus, the dynamic variables of the flow deviation and the variation of deviation are defined as in Table 5.3 where the absolute deviation of the air flow is used as the antecedent variable.

Figure 5.6 shows the control structure diagram of the SIRM-based dynamically connected fuzzy inference model. Four state variables are taken, which are the difference between the given value and the cathode pressure, and the change of the difference, the difference between the given value and the air flow, and the change of the difference, respectively. The absolute difference between the given value and cathode pressure is used as the antecedent variable of DID in pressure control loop and the absolute difference between the given value and air flow is considered as the antecedent variable of DID in flow control loop. The goal of optimization is to minimize the cumulative absolute values of the four state variables. Using simplified inference approach, the controls of the two loops can be achieved at the same time.

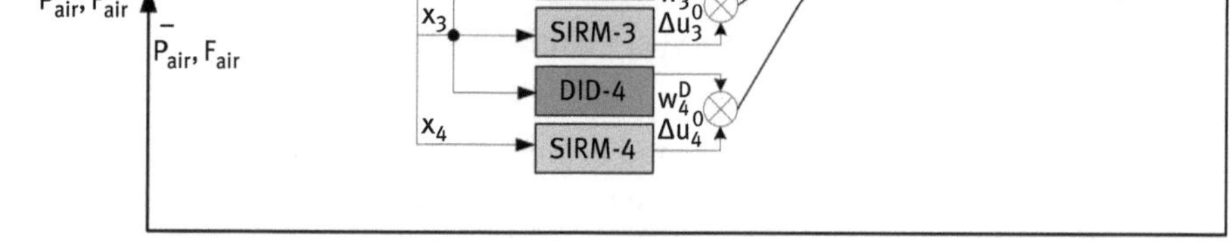

Figure 5.6: Structure diagram of the SIRM-based dynamically connected fuzzy controller.

5.3 Output tracking control

The hydrogen-air PEM fuel cell is a clean, quiet energy transformation device, which is very promising. However, the characteristic of the stack output displays a little bit "soft". It means that the produced electric current from the fuel cell cannot timely track the variations of the load under conventional control due to the high-order, nonlinear, and other uncertain features of fuel cell. This is why the fuel cell is usually formed as a hybrid system with battery or ultracapacitor to work. The battery or ultracapacitor can timely bridge the gap between the stack output and the demand. Serious deviation between the stack output and the actual demand will lead to the stack flooding or oxygen starvation, which will reduce the service life of the stack. The hybrid system forming with battery or ultracapacitor can cope with this kind of problem. However, it will increase the cost and complexity of the structure. In this section, we will try our best to design a new control strategy to realize the output tracking control of the high-order nonlinear hydrogen-air PEM fuel cell without the aid of a second energy device.

5.3.1 Fuzzy cluster modeling of fuel cell

The mechanism model of the fuel cell is too complex to realize output tracking control due to its high-order, nonlinear feature. Fortunately, fuzzy cluster modeling technology provides us a replacement, which can approximate a nonlinear system with some simple submodels by overlapping and smooth transitions of some membership functions [8, 9]. Then the output tracking control can be realized by sliding mode control technology. The model of the hydrogen-air PEM fuel cell can be described by a first-order discrete-time Nonlinear Auto-Regressive eXogenous (NARX) model:

$$y(k+1) = f(y(k), u(k)) \tag{5.7}$$

where k stands for the sampling instant and f is an unknown relationship which can be approximated by a Takagi-Sugeno (T-S) fuzzy model.

$$R_i\text{: } \textbf{If } y(k) \text{ is } A_i \textbf{ and } u(k) \text{ is } B_i$$

$$\textbf{then } y(k+1) = a_i y(k) + b_i u(k), \quad i = 1, 2, \cdots, K \tag{5.8}$$

where $y(k)$ is output, $u(k)$ is input, A_i and B_i are fuzzy sets which can be depicted by the membership functions from Gustafson-Kessel (GK) algorithm [10], and a_i, b_i can be identified by the least square estimation algorithm using the input-output data of the hydrogen-air PEM fuel cell.

The GK fuzzy clustering algorithm is an optimization problem of the cost function with the following form:

$$\min\left\{J(Z;U,V,A)=\sum_{i=1}^{c}\sum_{k=1}^{N}(\mu_{ik})^{m}D_{ikA_i}^{2}\right\}$$
$$D_{ikA_i}^{2}=(z_k-v_i)^{T}A_i(z_k-v_i) \tag{5.9}$$

where $Z=\{z_k|k=1,2,\cdots,N\}$ is an observation vector which is composed with the input-output data pairs from the controlled process. $A=[A_1,\cdots,A_c]$ is a set of c norm-inducing matrices, m is a fuzziness parameter of the clusters, $U=\mu_{ik}$ is a $c\times N$ matrix representing the fuzzy partition, $V=[V_1,\cdots,V_c]$ is a vector of cluster prototypes (centers), and $D_{ikA_i}^{2}$ is a squared inner-product distance norm. GK algorithm is an extension of the standard fuzzy c-means algorithm by employing an adaptive distance norm. It is realized by defining a norm-inducing matrix A_i for each cluster. Since unlimited A_i will affect the feasible solution of the optimization, a constraint is applied to the determinant of A_i.

$$|A_i|=\rho_i,\ \rho_i>0,\ \forall i \tag{5.10}$$

Using the Lagrange multiplier method, the following expression for A_i is presented:

$$A_i=[\rho_i\det(F_i)]^{1/n}F_i^{-1} \tag{5.11}$$

where F_i is the fuzzy covariance matrix of the ith cluster determined by

$$F_i=\frac{\sum_{k=1}^{N}\mu_{ik}^{m}(z_k-v_i)(z_k-v_i)^{T}}{\sum_{k=1}^{N}\mu_{ik}^{m}} \tag{5.12}$$

Then, the pointwise membership functions can be defined by

$$\mu_{ik}=\frac{1}{\sum_{j=1}^{c}(D_{ikA_i}/D_{jkA_i})^{2/(m-1)}} \tag{5.13}$$

To design a sliding mode controller, it will be very convenient for the pointwise membership functions to be translated into some mathematical functions. However, the pointwise membership functions are usually replaced by the piecewise exponential membership functions. The remarkable features of these alternative membership functions are indirectly determined by their implicit antecedent variables $y(k)$ or $u(k)$ and cannot be directly utilized in the controller design. To solve this kind of problem, two-layer feedforward networks with sigmoid hidden neurons and linear output neurons are utilized to transform the implicit form of membership functions into the explicit form of membership functions (see Figure 5.7). The process can be described by the following equation:

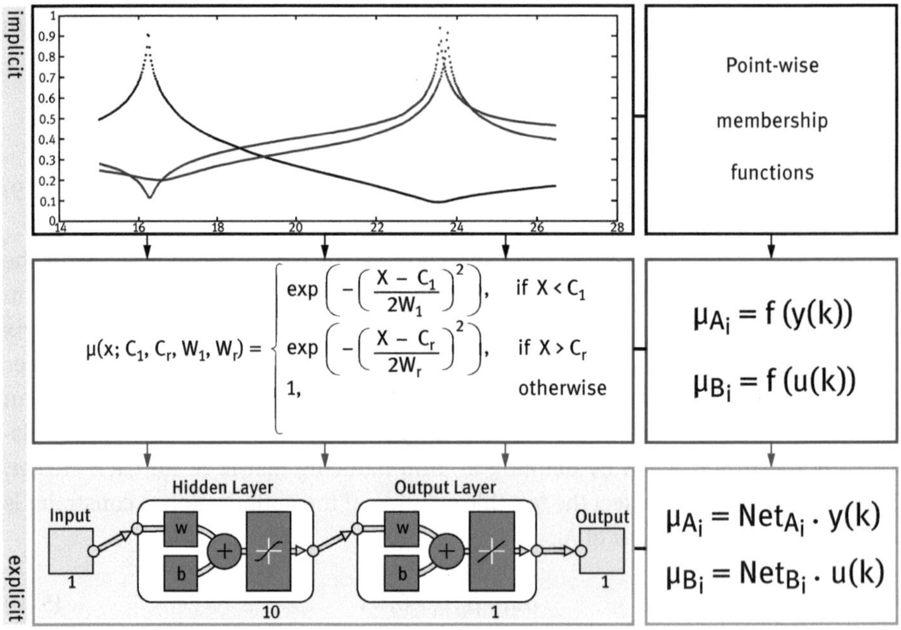

Figure 5.7: Transforming the implicit membership functions into the explicit membership functions.

$$\left.\begin{aligned}\mu_{A_i} &= f(y(k))\\ \mu_{B_i} &= f(u(k))\end{aligned}\right\} \Rightarrow \left.\begin{aligned}\mu_{A_i} &= \mathrm{Net}_{A_i}\cdot y(k)\\ \mu_{B_i} &= \mathrm{Net}_{B_i}\cdot u(k)\end{aligned}\right\} \tag{5.14}$$

Thus, the inference results of the T-S fuzzy model can be easily obtained by

$$\begin{aligned} y(k+1) &= \frac{\sum_{i=1}^{k}\mu_{A_i}\cdot\mu_{B_i}(a_i y(k)+b_i u(k))}{\sum_{i=1}^{k}\mu_{A_i}\cdot\mu_{B_i}}\\ &= \frac{\sum_{i=1}^{k}(\mathrm{Net}_{A_i}\cdot y(k))(\mathrm{Net}_{B_i}\cdot u(k))(a_i y(k)+b_i u(k))}{\sum_{i=1}^{k}(\mathrm{Net}_{A_i}\cdot y(k))(\mathrm{Net}_{B_i}\cdot u(k))}\\ &= \frac{\sum_{i=1}^{k}\mathrm{Net}_{A_i}\cdot\mathrm{Net}_{B_i}(a_i y(k)+b_i u(k))}{\sum_{i=1}^{k}\mathrm{Net}_{A_i}\cdot\mathrm{Net}_{B_i}} \end{aligned} \tag{5.15}$$

and the estimation of the consequent parameters a_i and b_i can be regarded as the least square estimation issue with the formula

$$Y = X\cdot\theta + \varepsilon \tag{5.16}$$

where $Y = [y(k+1), y(k+2), \cdots, y(k+N)]^T$ is the output matrix, $\theta = [a_1, b_1, a_2, b_2, \cdots, a_c, b_c]^T$ is the parameters matrix, and ε is the noise matrix. The measurement matrix X is written by

$$X = \begin{bmatrix} [y(k)\ u(k)] & [y(k)\ u(k)] & \cdots & [y(k)\ u(k)] \\ [y(k+1) & [y(k+1) & \cdots & [y(k+1) \\ u(k+1)] & u(k+1)] & & u(k+1)] \\ \vdots & \vdots & \ddots & \vdots \\ [y(k+N-1) & [y(k+N-1) & \cdots & [y(k+N-1) \\ u(k+N-1)] & u(k+N-1)] & & u(k+N-1)] \end{bmatrix}_{N\times c} \cdot W$$

$$W = \frac{1}{\mu_{A_1}\mu_{B_1} + \mu_{A_2}\mu_{B_2} + \cdots + \mu_{A_c}\mu_{B_c}} \begin{bmatrix} \mu_{A_1}\mu_{B_1} & & & \\ & \mu_{A_2}\mu_{B_2} & & \\ & & \ddots & \\ & & & \mu_{A_c}\mu_{B_c} \end{bmatrix}_{c\times c} .$$

Then, the answer of the least square estimation problem is written by

$$\theta = (X^TX)^{-1}X^TY \tag{5.17}$$

5.3.2 Sliding mode output tracking control

It can be seen from equation (5.15), $y(k+1)$ becomes an explicit function of variables $y(k)$ and $u(k)$ after replacing the pointwise membership functions with the neural networks. Rewritten equation (5.15) yields

$$y(k+1) = \hat{a}y(k) + \hat{b}u(k) \tag{5.18}$$

where

$$\hat{a} = \left(\sum_{i=1}^{k} Net_{A_i} \cdot Net_{B_i} \cdot a_i \Big/ \sum_{i=1}^{k} Net_{A_i} \cdot Net_{B_i}\right),$$

$$\hat{b} = \left(\sum_{i=1}^{k} Net_{A_i} \cdot Net_{B_i} \cdot b_i \Big/ \sum_{i=1}^{k} Net_{A_i} \cdot Net_{B_i}\right).$$

Let T be the sample time, $\tilde{y}(k+1) = (y(k+1) - y(k))/T$, $\tilde{y}(k) = (y(k) - y(k-1))/T$; we can obtain the equation

$$\tilde{y}(k+1)=\hat{a}\tilde{y}(k)+\frac{\hat{b}}{T}u(k)-\frac{\hat{b}}{T}u(k-1) \tag{5.19}$$

Define error $e(k)=r(k)-y(k)$, $e(k+1)=r(k+1)-y(k+1)$, change rate of error $\tilde{e}(k)=\tilde{r}(k)-\tilde{y}(k)$, $\tilde{e}(k+1)=\tilde{r}(k+1)-\tilde{y}(k+1)$, reference $r(k+1)=2r(k)-r(k-1)$ and change rate of reference $\tilde{r}(k+1)=2\tilde{r}(k)-\tilde{r}(k-1)$ at sample time, respectively.

The switch functions at sample time k and $k+1$ can be defined by

$$s(k)=ce(k)+\tilde{e}(k)=c(r(k)-y(k))+\tilde{r}(k)-\tilde{y}(k) \tag{5.20}$$

and

$$\begin{aligned} s(k+1)&=ce(k+1)+\tilde{e}(k+1)=c(r(k+1)-y(k+1))+\tilde{r}(k+1)-\tilde{y}(k+1)\\ &=c\left(r(k+1)-\hat{a}y(k)-\hat{b}u(k)\right)+\tilde{r}(k+1)-\hat{a}\tilde{y}(k)-\frac{\hat{b}}{T}u(k)+\frac{\hat{b}}{T}u(k-1) \end{aligned} \tag{5.21}$$

From equation (5.21), the control action $u(k)$ is derived by

$$u(k)=\frac{1}{Tc\hat{b}+\hat{b}}\left(cTr(k+1)-cT\hat{a}y(k)+T\tilde{r}(k+1)-\hat{a}T\tilde{y}(k)+\hat{b}u(k-1)-Ts(k+1)\right) \tag{5.22}$$

Design a discrete form of reaching law

$$s(k+1)=s(k)+T(-\varepsilon\cdot\operatorname{sgn}(s(k))-q\cdot s(k)) \tag{5.23}$$

Substituting (5.23) for (5.22), the control law is obtained.

$$u(k)=\frac{1}{Tc\hat{b}+\hat{b}}\left(cTr(k+1)-cT\hat{a}y(k)+T\tilde{r}(k+1)-\hat{a}T\tilde{y}(k)+\hat{b}u(k-1)-Ts(k)-T\tilde{s}(k)\right) \tag{5.24}$$

where $\tilde{s}(k)=T(-\in\cdot\operatorname{sgn}(s(k))-q\cdot s(k))$.

5.3.3 Steady-state compensation

Although the fuzzy clustering model can basically reflect the dynamic response of the nonlinear fuel cell system, it cannot completely take the place of the fuel cell system due to some uncertainties. However, the insufficiency can be made up by importing an integrative control action. The kernel idea is to track the dynamic response with the fuzzy clustering model and to eliminate the steady-state error with the integrative control action. The integrative controller is an auxiliary control means, which can be designed by trial and error. The most important criterion for the setting of integrative parameter is to keep the stability of the control system.

5.4 Simulations and experiments

5.4.1 Fuzzy control based on SIRMs dynamically connected fuzzy inference model and the random search

1) Simulations

The model of the 1kW hydrogen-air PEM fuel cell in the Institute of Automation, Chinese Academy of Sciences, in the Simulink environment can be built according to the mass conservation, ideal gas equation, Dalton's law of partial pressure, and electrochemical and other mechanism analysis. Part of the model parameters comes from references (such as stack voltage models), partially adjusted according to the experimental curve; others come from the experimental data fitting (eg, gas flow equations of the cathode and the anode outlet path) or identification (eg, gas flow model of the cathode and the anode inlet path). Figure 5.8 is a configuration diagram for the simulative control of the fuel cell flow underactuated system based on the SIRMs dynamically connected fuzzy controller where the cathode pressure is controlled at 30 kPa, the anode pressure follows the cathode pressure changes, and the set value of the oxygen flow can be calculated according to the electrochemical principle by the feedback current of the load. The stack works on 65°C temperature. The module "Controller based on the SIRMs Fuzzy Inference Model" is written in C-S function with two versions: one is fixed dynamic importance degree and the other is online optimization dynamic importance degree. The fuel cell output voltage is reduced by the DC/DC converter to a standard 12V. The fuel cell model is derived from electrochemical principles, mass conservation, ideal gas equations, and so on, and some of the parameters are from the actual device. Figure 5.9 is an off-line optimized simulative result, and Figure 5.10 is an online optimized simulative result. As can be seen from the simulation results, the fuzzy control method based on SIRMs dynamically connected fuzzy inference model can realize the simultaneous control requirement of the cathode pressure and the air flow for the hydrogen-air fuel cell flow system.

2) Experiment

In the Simulink environment, a fuzzy control method for the hydrogen-air PEM fuel cell flow underactuated system based on the SIRMs dynamically connected

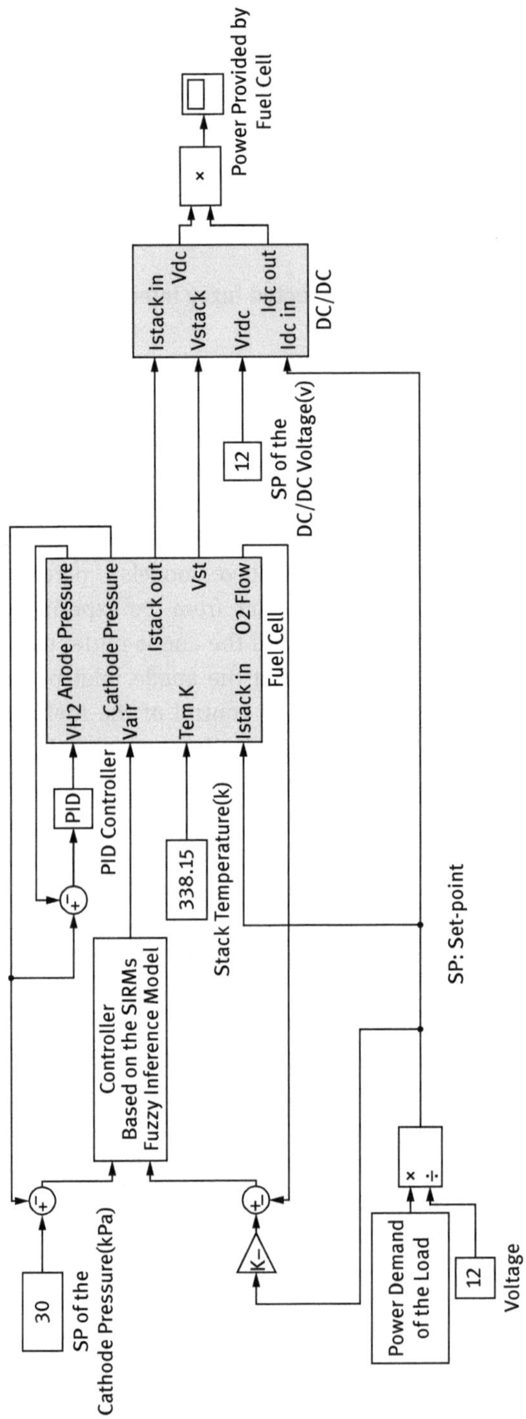

Figure 5.8: Simulative control configuration diagram for the PEM fue cell flow underactuated system in Simulink environment.

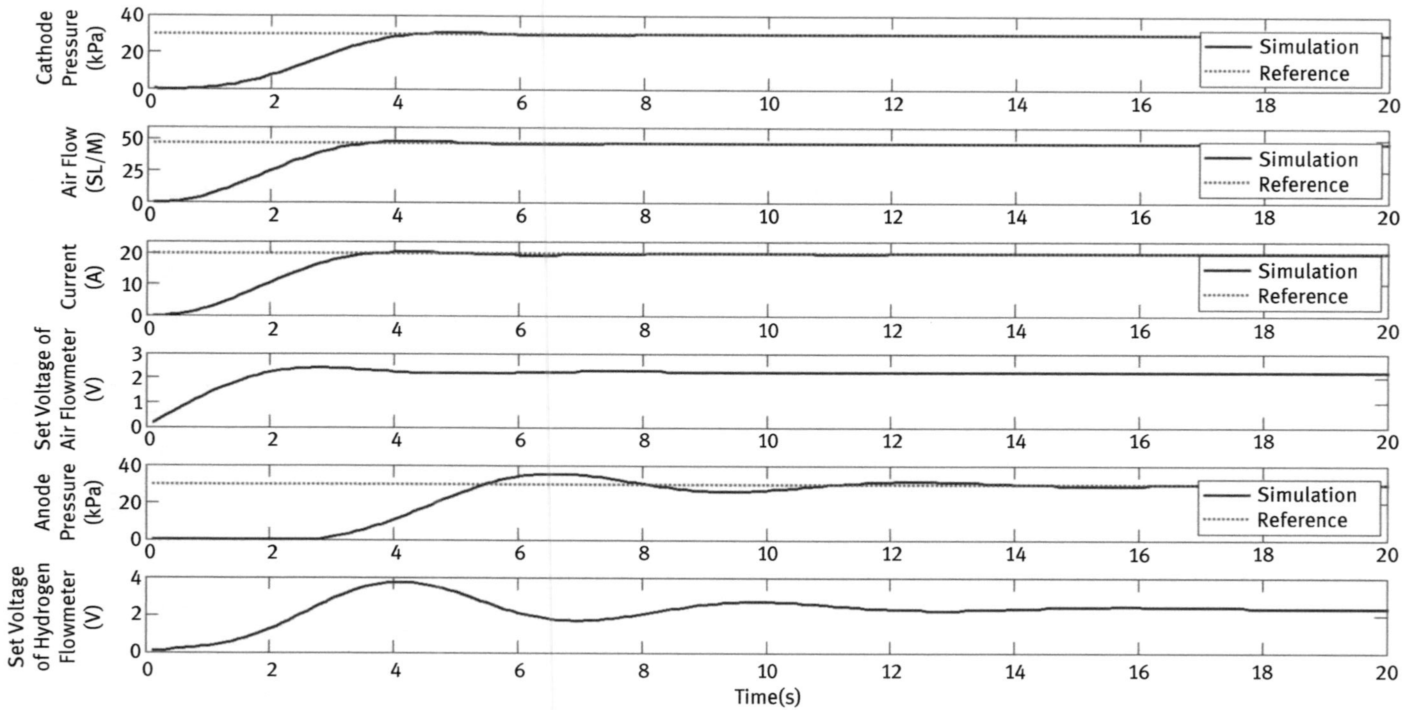

Figure 5.9: Simulative result for the controller based on the SIRMs dynamically connected fuzzy inference model (off-line optimization).

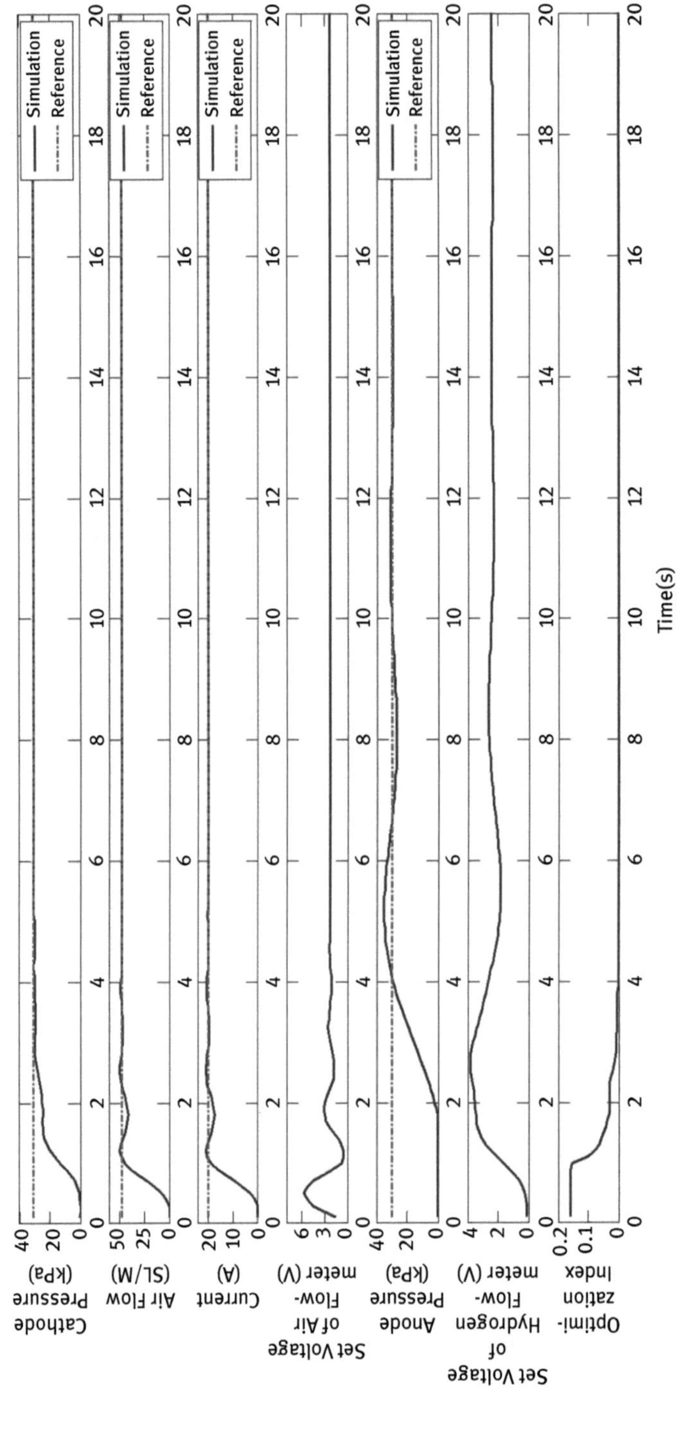

Figure 5.10: Simulative result for the controller based on the SIRMs dynamically connected fuzzy inference model (online optimization).

fuzzy inference model is shown in Figure 5.11. Different from simulation, the control configuration needs to be converted into executable code to run in the controller. The cathode pressure, air flow, load current, anode pressure, and other analog are first collected from the actual process, and then converted into digital. After the operation of the fuzzy controller, the digital control actions are obtained, which is applied to the fuel cell device through DC/AC conversion to achieve the control of the fuel cell flow underactuated system. Figure 5.12 is the control performance using the fixed dynamic importance degrees while Figure 5.13 is the control performance using online optimization of dynamic importance degrees; although there are fluctuations in the steady state, the demand to control two variables with one means is basically achieved.

5.4.2 Output tracking control

To design the sliding mode controller for the 1kW hydrogen-air PEM fuel cell system, the mechanism model is not convenient to be directly utilized due to the high-order and nonlinear characteristics. However, an alternative T-S model can be obtained from the input-output data of the process by using the fuzzy clustering technology. The multi-sinusoidal signals with 25 amplitude, $0.06\pi, 0.012\pi, 0.03\pi$ rad/sec frequency are selected as exciting sources to add to the set voltage of the air mass flowmeter (hydrogen mass flowmeter) under the variations of the stack current. These input signals and the stack output voltages are collected as the data set for modeling. They are divided into two halves. One half is for fuzzy cluster modeling, and the other half is for validation. The number of clusters is defined by $c = 3$, the fuzziness parameter is denoted by $m = 2$, and the termination criterion is 0.01. From Figure 5.14, it can be seen that the output of the fuzzy clustering model (magenta color) agrees very well with the validation data (blue color) from the process (see the top subgraph). This is done by the overlap of three individual local models (see the middle subgraph) with smooth transitions from the definition of the membership functions (see the bottom sub-graph).

To design a sliding mode controller, it is better to formulate the explicit membership functions for the input variable $u(k)$ and the output variable $y(k)$ since the input and output variables that simultaneously appeared in the numerator and denominator can be canceled. Thus, the control law for the sliding mode controller can be calculated directly. In this book, two-layer feedforward networks with ten sigmoid hidden neurons and linear output neurons are used to fit the pointwise membership functions generated by the fuzzy clustering technology and then construct explicit functions for the input and output variables.

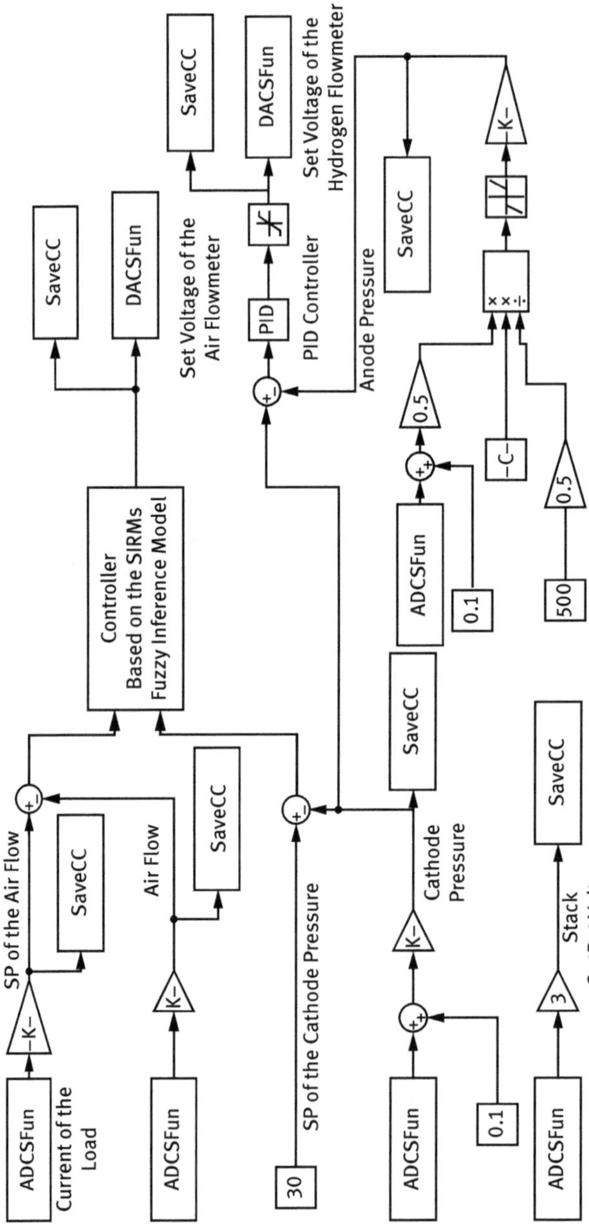

Figure 5.11: Actual control configuration diagram for the PEM fuel cell flow underactuated system in Simulink environment.

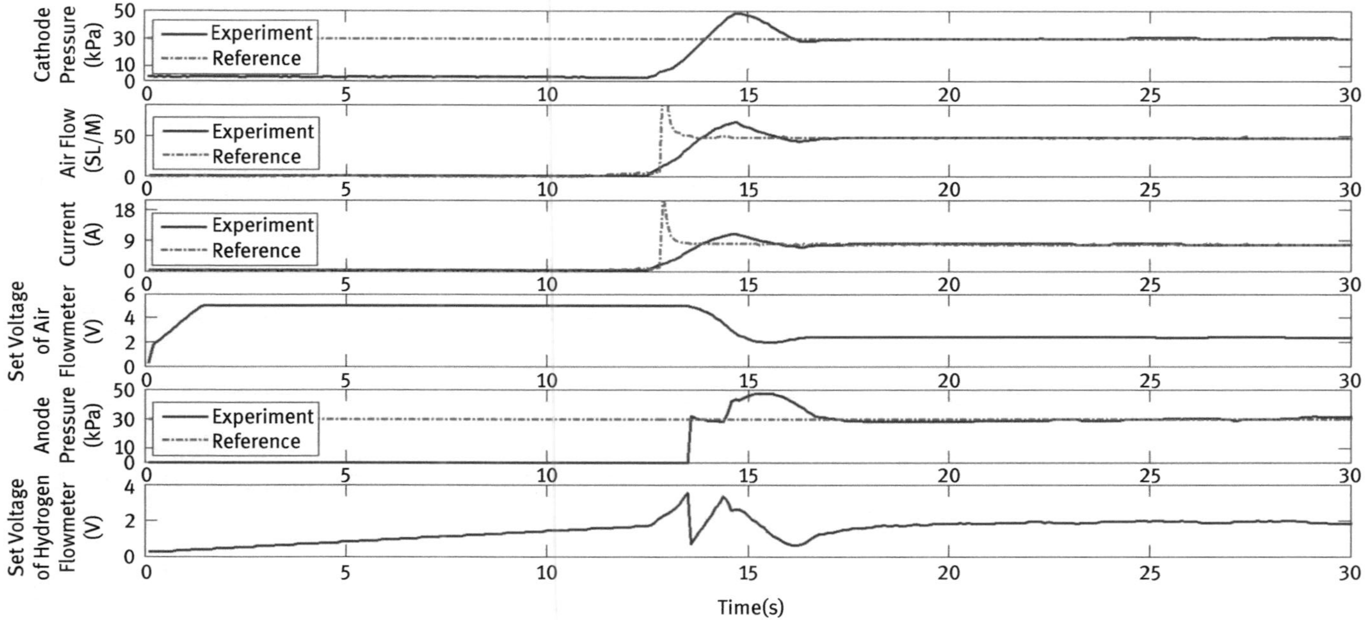

Figure 5.12: Experimental result for the controller based on the SIRMs dynamically connected fuzzy inference model (off-line optimization).

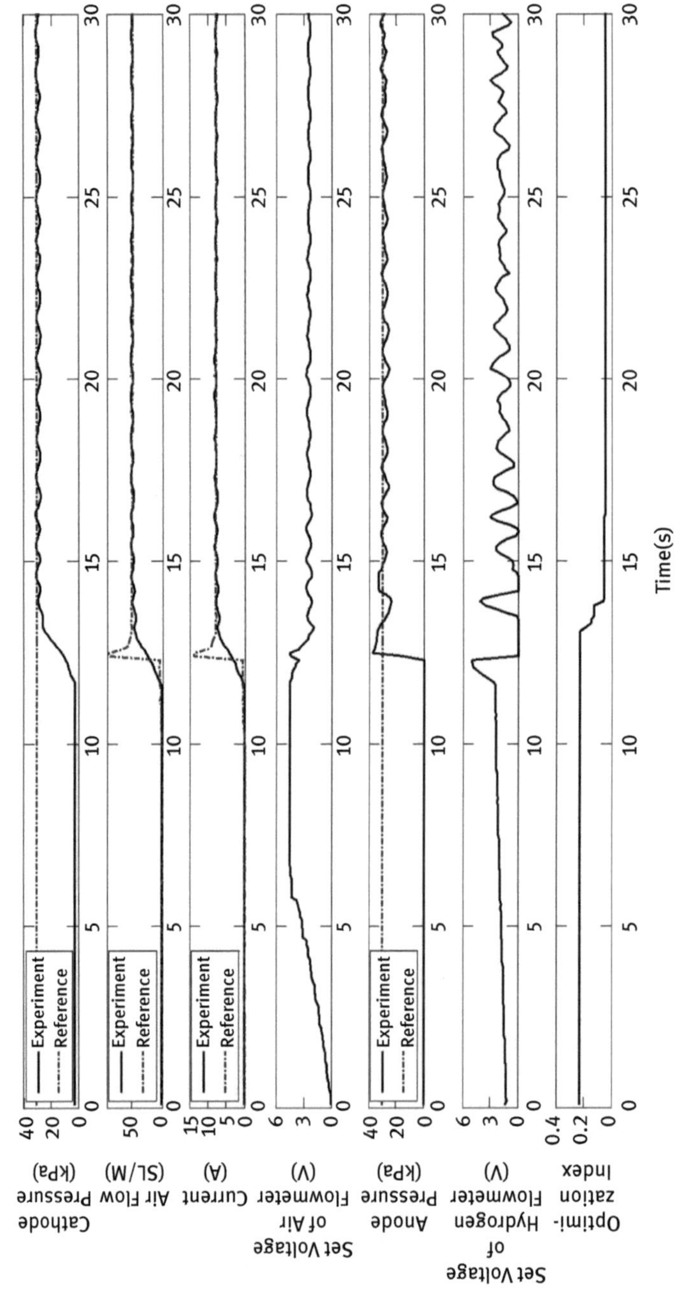

Figure 5.13: Experimental result for the controller based on the SIRMs dynamically connected fuzzy inference model (online optimization).

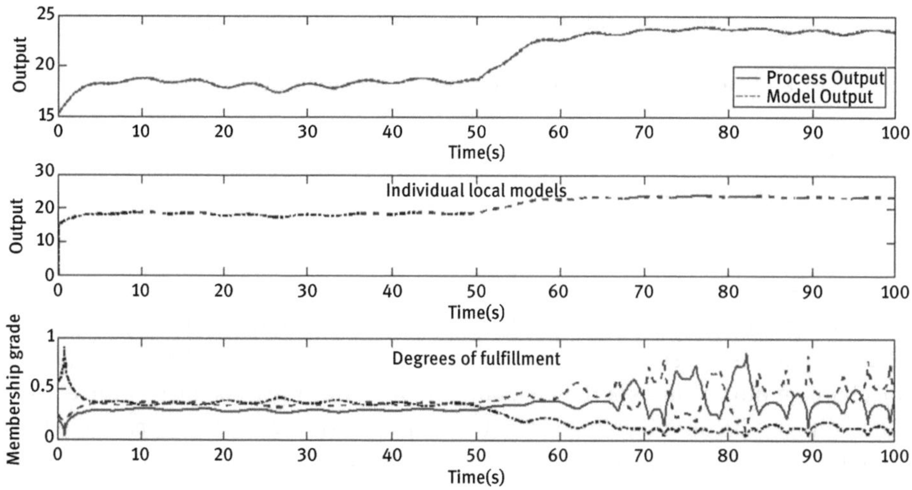

Figure 5.14: Comparison of the agreement between the process output and the model output, individual local models, and the degrees of fulfillment.

The three individual local models are constructed by

$$\mathbf{R_1: If}\ y(k)\ \text{is}\ A_1\ \mathbf{and}\ u(k)\ \text{is}\ B_1$$

$$\mathbf{then}\ y(k+1) = 0.9892y(k) + 0.1033u(k)$$

$$\mathbf{R_2: If}\ y(k)\ \text{is}\ A_2\ \mathbf{and}\ u(k)\ \text{is}\ B_2$$

$$\mathbf{then}\ y(k+1) = 1.0159y(k) - 0.2049u(k)$$

$$\mathbf{R_3: If}\ y(k)\ \text{is}\ A_3\ \mathbf{and}\ u(k)\ \text{is}\ B_3$$

$$\mathbf{then}\ y(k+1) = 0.9906y(k) + 0.1110u(k) \tag{5.25}$$

According to equation (5.15), equation (5.25), and the calculation results from the neural networks, the substituting model for the hydrogen-air PEM fuel cell is written by

$$y(k+1) = 0.8602y(k) + 1.6319u(k) \tag{5.26}$$

Define the second equation using the incremental variables of the output $y(k)$ and $y(k+1)$

$$\tilde{y}(k+1) = 0.8602\tilde{y}(k) + 1.6319u(k) - 1.6319u(k-1) \tag{5.27}$$

Let $c = 30$, $\varepsilon = 5$, and $q = 50$; the control action is obtained by:

$$\begin{aligned} u(k) = {} & 0.4596r(k+1) - 0.3953y(k) + 0.01532\tilde{r}(k+1) - 0.01318\tilde{y}(k) + 0.25u(k-1) \\ & - 0.01532s(k) - 0.01532\tilde{s}(k) \end{aligned} \tag{5.28}$$

where $\tilde{s}(k) = -0.5\mathrm{sgn}(s(k)) - 5s(k)$. Then the simulation is configured in the Simulink environment as shown in Figure 5.15. The model of the hydrogen-air PEM fuel cell is constructed using the blocks in the Simulink library, which is a mechanism plus semi-experience model. It simulates the dynamics of the activation voltage, the ohmic voltage loss, the concentration voltage, the cathode flow and the anode flow. The reference stack voltage is determined by the polarization curve (it is a nonlinear relationship) with stack output current as input which

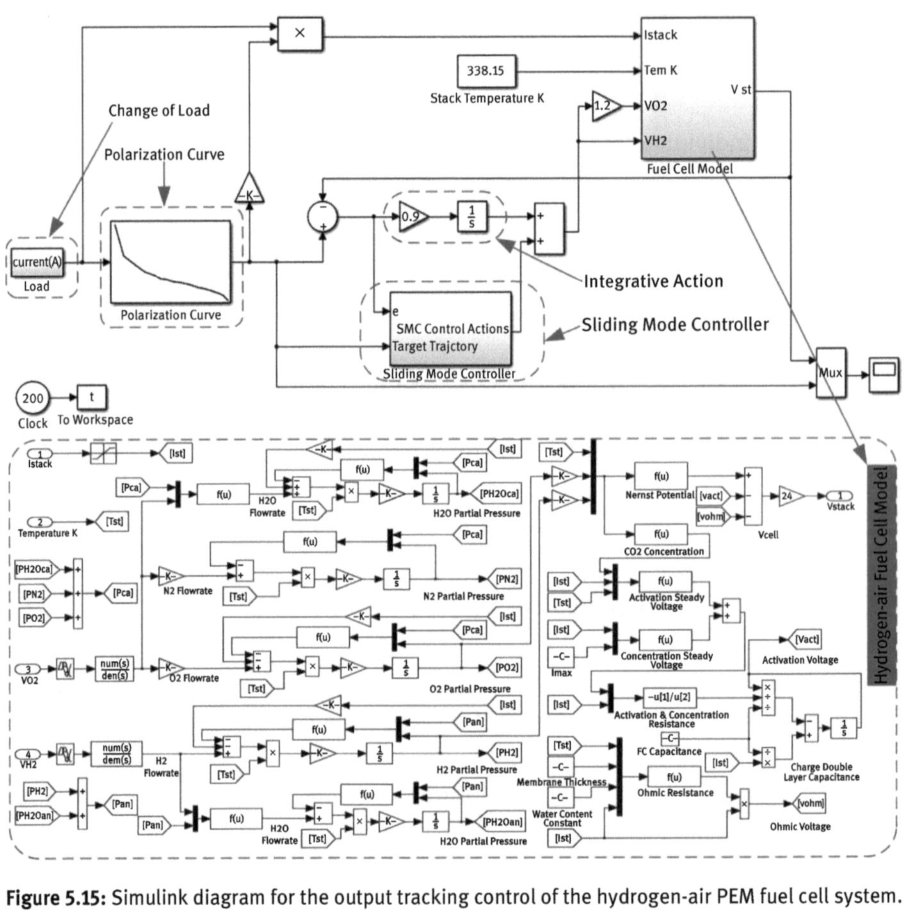

Figure 5.15: Simulink diagram for the output tracking control of the hydrogen-air PEM fuel cell system.

represents the variations of the load. The stack temperature is controlled at 65°C and the stack output voltage is regulated by setting the set voltage of the air mass flowmeter or the hydrogen mass flowmeter. Control actions are composed with two parts: one is the output of a sliding mode controller that is used to track the dynamic response of the fuel cell, and the other one is from an integrative controller which is used to compensate for the steady-state error (the integrative coefficient is 0.9 in the simulation). The sliding mode controller is written in Matlab m-sfunction.

The control performance is presented in Figure 5.16. It can be seen that the stack output voltage follows very well with the reference signal (see top subgraph) which is polarized by the current of the load (see middle subgraph). Control actions are displayed in the bottom subgraph. The dash-dot line in red color is the sliding mode controller output, the dash line in green color is the integrative compensation output, and the solid line in blue color is the compound signal which is used as the set voltage of the air mass flowmeter or the hydrogen mass flowmeter. It can be seen that the control actions generated from the compound controller can be physically realized since the variations of the voltage is between 0V and 5V. Without the aid of a second energy source, the control algorithm proposed in this chapter can achieve the output tracking control of the high-order, nonlinear hydrogen-air PEM fuel cell system.

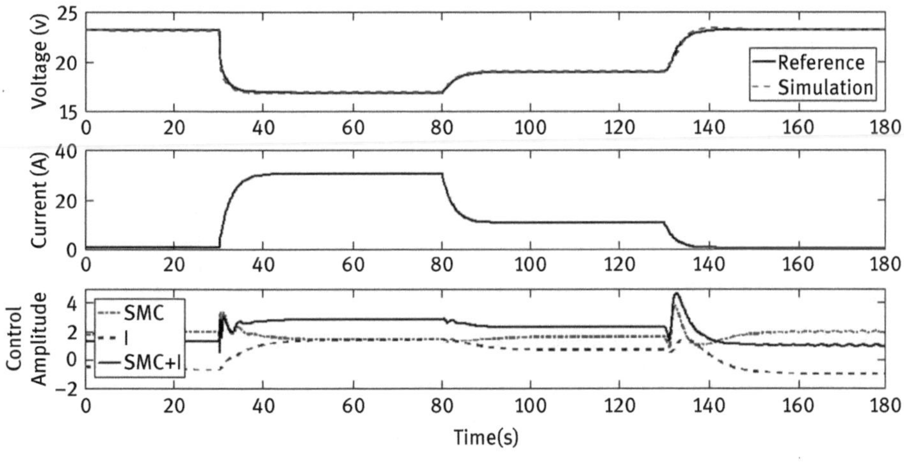

Figure 5.16: Simulative results for the output tracking control of the hydrogen-air PEM fuel cell system.

To validate the proposed method, we compare it with a PID controller. The PID controller parameters are set by $K_p = 1.5$, $K_i = 1.8$, $K_d = 0.01$. Figure 5.17 shows the simulative results. It can be seen that the PID controller works very well between the

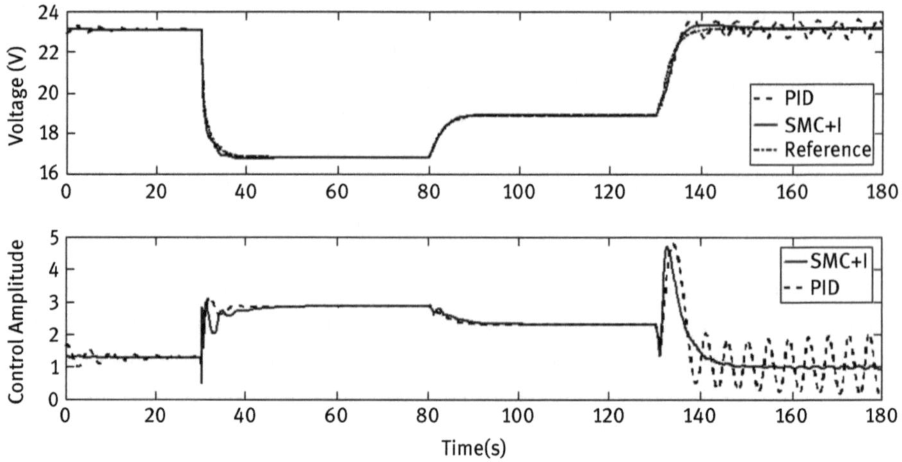

Figure 5.17: Control performance comparison with PID controller.

30th second and the 130th second. However, it begins to oscillate after the 130th second due to the nonlinear characteristic of the fuel cell. In contrast, our proposed method still shows good performance.

5.5 Source codes

5.5.1 SIRMs-based dynamically connected fuzzy inference control (Fixed ID)

1) Simulink diagram of the SIRMs-based dynamically connected fuzzy inference control (Fixed ID)

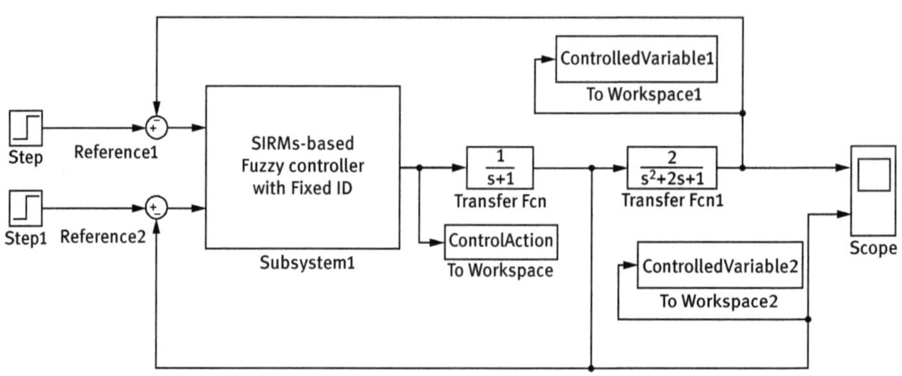

Figure 5.18: Simulink diagram of SIRMs-based fuzzy control with fixed ID.

Note: For better understanding the source codes, some examples are presented. Please see Figure 5.18–5.31.

2) Simulink diagram of the SIRMs-based fuzzy controller subsystem with fixed ID

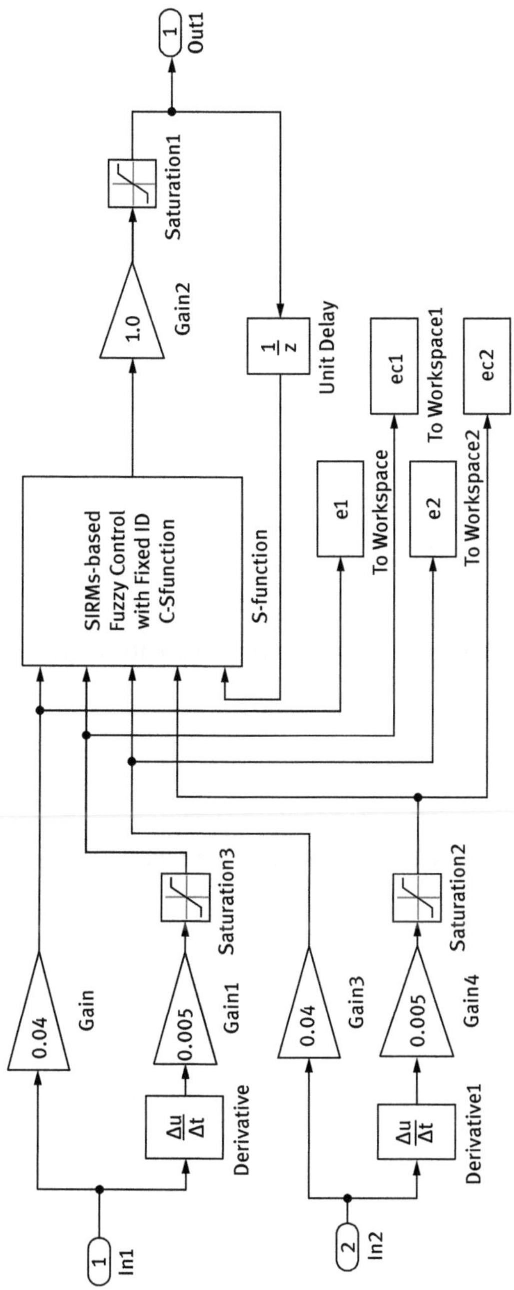

Figure 5.19: Simulink diagram of SIRMs-based fuzzy controller subsystem with fixed ID.

3) C-Sfunction for the SIRMs-based dynamically connected fuzzy inference control (Fixed ID)

```
/*C-Sfunction: sfSIRMFuzzyConFixedID.c==============================
 * Abstract:
 *  This C-S function is used to simulate the SIRMs-based fuzzy control
    algorithm
 *  with fixed ID.
 */
 #define S_FUNCTION_NAME sfSIRMFuzzyConFixedID
 #define S_FUNCTION_LEVEL 2
/*
 * Need to include simstruc.h for the definition of the SimStruct and
 * its associated macro definitions.
 */
 #include "simstruc.h"
 #include "math.h"

 #define sampletime(S) ssGetSFcnParam(S, 0)
/* Error handling
 *
 * You should use the following technique to report errors encountered within
 * an S-function:
 *
 *    ssSetErrorStatus(S,"Error encountered due to ...");
 *    return;
 *
 *  Note that the 2nd argument to ssSetErrorStatus must be persistent
    memory.
 * It cannot be a local variable. For example the following will cause
 * unpredictable errors:
 *
 *   mdlOutputs()
 *   {
 *     char msg[256];      {ILLEGAL: to fix use "static char msg[256];"}
 *     sprintf(msg,"Error due to %s", string);
 *     ssSetErrorStatus(S,msg);
 *     return;
 *   }
 *
 * See matlabroot/simulink/src/sfuntmpl_doc.c for more details.
 */
```

```
/*====================*
 * S-function methods *
 *====================*/

/*Function: mdlInitializeSizes ========================================
 * Abstract:
 *  The sizes information is used by Simulink to determine the S-function
 *  block's characteristics (number of inputs, outputs, states, etc.).
 */
 static int_T count;          //Counting variable used to set initial value
 static void mdlInitializeSizes(SimStruct *S)
 {
   /* See sfuntmpl_doc.c for more details on the macros below */
   ssSetNumSFcnParams(S, 1); /* Number of expected parameters */
   if (ssGetNumSFcnParams(S) != ssGetSFcnParamsCount(S)) {
       /* Return if number of expected != number of actual parameters */
       return;
   }

   ssSetNumContStates(S, 0);
   ssSetNumDiscStates(S, 0);

   if (!ssSetNumInputPorts(S, 5)) return;  //Specify the number of input
ports that a block has.
   ssSetInputPortWidth(S, 0, 1);       //Specify the width of an input port.
   ssSetInputPortWidth(S, 1, 1);       //Specify the width of an input port.
   ssSetInputPortWidth(S, 2, 1);       //Specify the width of an input port.
   ssSetInputPortWidth(S, 3, 1);       //Specify the width of an input port.
   ssSetInputPortWidth(S, 4, 1);       //Specify the width of an input port.
  ssSetInputPortRequiredContiguous(S, 0, true); /*direct input signal
access*/
   ssSetInputPortRequiredContiguous(S, 1, true); /*direct input signal
access*/
   ssSetInputPortRequiredContiguous(S, 2, true); /*direct input signal
access*/
   ssSetInputPortRequiredContiguous(S, 3, true); /*direct input signal
access*/
   ssSetInputPortRequiredContiguous(S, 4, true); /*direct input signal
access*/
  /*
   * Set direct feedthrough flag (1=yes, 0=no).
   * A port has direct feedthrough if the input is used in either
   * the mdlOutputs or mdlGetTimeOfNextVarHit functions.
```

```
    * See matlabroot/simulink/src/sfuntmpl_directfeed.txt.
    */
   ssSetInputPortDirectFeedThrough(S, 0, 1); //Specify the direct feed-
through status of a block's ports.
   ssSetInputPortDirectFeedThrough(S, 1, 1); //Specify the direct feed-
through status of a block's ports.
   ssSetInputPortDirectFeedThrough(S, 2, 1); //Specify the direct feed-
through status of a block's ports.
   ssSetInputPortDirectFeedThrough(S, 3, 1); //Specify the direct feed-
through status of a block's ports.
   ssSetInputPortDirectFeedThrough(S, 4, 1); //Specify the direct feed-
through status of a block's ports.

   if (!ssSetNumOutputPorts(S, 1)) return;
   ssSetOutputPortWidth(S, 0, 1);

   ssSetNumSampleTimes(S, 1);      //Specify the number of sample times that
an S-Function block has.
   ssSetNumRWork(S, 0);
   ssSetNumIWork(S, 0);
   ssSetNumPWork(S, 0);
   ssSetNumModes(S, 0);
   ssSetNumNonsampledZCs(S, 0);

   ssSetOptions(S, 0);
  }

 /*Function: mdlInitializeSampleTimes ================================
  * Abstract:
  *  This function is used to specify the sample time(s) for your
  *  S-function. You must register the same number of sample times as
  *  specified in ssSetNumSampleTimes.
  */
  static void mdlInitializeSampleTimes(SimStruct *S)
  {
   const real_T  *stime=mxGetPr(sampletime(S));
   ssSetSampleTime(S, 0, stime[0]);
   ssSetOffsetTime(S, 0, 0.0);
  }

#define MDL_INITIALIZE_CONDITIONS /* Change to #undef to remove function
*/
#if defined(MDL_INITIALIZE_CONDITIONS)
```

```
/*Function: mdlInitializeConditions ==================================
 * Abstract:
 *  In this function, you should initialize the continuous and discrete
 *  states for your S-function block. The initial states are placed
 *  in the state vector, ssGetContStates(S) or ssGetRealDiscStates(S).
 *  You can also perform any other initialization activities that your
 *  S-function may require. Note, this routine will be called at the
 *  start of simulation and if it is present in an enabled subsystem
 *  configured to reset states, it will be call when the enabled subsystem
 *  restarts execution to reset the states.
 */
 static void mdlInitializeConditions(SimStruct *S)
 {
 }
 #endif /* MDL_INITIALIZE_CONDITIONS */

 #define MDL_START /* Change to #undef to remove function */
 #if defined(MDL_START)
/*Function: mdlStart ===============================================
 * Abstract:
 *  This function is called once at start of model execution. If you
 *  have states that should be initialized once, this is the place
 *  to do it.
 */
 static void mdlStart(SimStruct *S)
 {
   count=0;
 }
 #endif /* MDL_START */

/*Function: mdlOutputs =============================================
 * Abstract:
 *  In this function, you compute the outputs of your S-function
 *  block. Generally outputs are placed in the output vector, ssGetY(S).
 */

 static void mdlOutputs(SimStruct *S, int_T tid)
 {
   const real_T  *p = ssGetInputPortRealSignal(S,0);
   const real_T  *dp = ssGetInputPortRealSignal(S,1);
   const real_T  *v = ssGetInputPortRealSignal(S,2);
   const real_T  *dv = ssGetInputPortRealSignal(S,3);
```

```
const real_T  *lu = ssGetInputPortRealSignal(S,4);
const real_T  *stime = mxGetPr(sampletime(S));
real_T     *u = (real_T *)ssGetOutputPortSignal(S,0);
real_T Set_A[7] = {-1,-1,0,0,0,1,1};
real_T Set_B[7] = {-1,-1,0,0,0,1,1};
real_T Set_C[7] = {-1,-1,0,0,0,1,1};
real_T Set_D[7] = {-1,-1,0,0,0,1,1};
real_T Set_E[7] = {0,0,0.5,0.5,0.5,1,1};
real_T Set_F[7] = {0,0,0.5,0.5,0.5,1,1};
real_T A_p[3]  = {0,0,0};
real_T A_dp[3] = {0,0,0};
real_T A_v[3]  = {0,0,0};
real_T A_dv[3] = {0,0,0};
real_T A_ap[3] = {0,0,0};
real_T A_av[3] = {0,0,0};
real_T du_p[3] = {-1,0,1};
real_T du_dp[3] = {-1,0,1};
real_T du_v[3] = {-1,0,1};
real_T du_dv[3] = {-1,0,1};
real_T dw_p[3] = {0,0.5,1};
real_T dw_dp[3] = {0,0.5,1};
real_T dw_v[3] = {0,0.5,1};
real_T dw_dv[3] = {0,0.5,1};
real_T wB[4]  = {0,0,0,0};
real_T B[4]  = {1,1,1,1};
real_T du0[4],dw0[4],du,wD[4];
real_T tmp_p,tmp_dp,tmp_v,tmp_dv,tmp_ap,tmp_av,sumdu;
real_T sumdu1[4],sumdu2[4],sumdw1[4],sumdw2[4],tmp_A1[3];
real_T tmp_du[3],tmp_A2[3],tmp_dw[3];
int_T i,j,m;

tmp_p = p[0];
tmp_dp = dp[0];
tmp_v = v[0];
tmp_dv = dv[0];
tmp_ap = fabs(p[0]);
tmp_av = fabs(v[0]);

////// The First Table: p //////
if (tmp_p<Set_A[0])
A_p[0]=1;
else if (tmp_p>=Set_A[0]&&tmp_p<=Set_A[2])
```

```
  A_p[0]=-1/(Set_A[2]-Set_A[0])*(tmp_p-Set_A[0])+1;
  else;
  if (tmp_p>=Set_A[1]&&tmp_p<Set_A[3])
  A_p[1]=1/(Set_A[3]-Set_A[1])*(tmp_p-Set_A[3])+1;
  else if (tmp_p>=Set_A[3]&&tmp_p<Set_A[5])
  A_p[1]=-1/(Set_A[5]-Set_A[3])*(tmp_p-Set_A[3])+1;
  else;
  if (tmp_p>=Set_A[4]&&tmp_p<Set_A[6])
  A_p[2]=1/(Set_A[6]-Set_A[4])*(tmp_p-Set_A[6])+1;
  else if (tmp_p>=Set_A[6])
  A_p[2]=1;
  else;
////// The End of The First Table //////

////// The Second Table: dp //////
  if (tmp_dp<Set_B[0])
  A_dp[0]=1;
  else if (tmp_dp>=Set_B[0]&&tmp_dp<=Set_B[2])
  A_dp[0]=-1/(Set_B[2]-Set_B[0])*(tmp_dp-Set_B[0])+1;
  else;
  if (tmp_dp>=Set_B[1]&&tmp_dp<Set_B[3])
  A_dp[1]=1/(Set_B[3]-Set_B[1])*(tmp_dp-Set_B[3])+1;
  else if (tmp_dp>=Set_B[3]&&tmp_dp<Set_B[5])
  A_dp[1]=-1/(Set_B[5]-Set_B[3])*(tmp_dp-Set_B[3])+1;
  else;
  if (tmp_dp>=Set_B[4]&&tmp_dp<Set_B[6])
  A_dp[2]=1/(Set_B[6]-Set_B[4])*(tmp_dp-Set_B[6])+1;
  else if (tmp_dp>=Set_B[6])
  A_dp[2]=1;
  else;
////// The End of The Second Table //////

////// The Third Table: v //////
  if (tmp_v<Set_C[0])
  A_v[0]=1;
  else if (tmp_v>=Set_C[0]&&tmp_v<=Set_C[2])
  A_v[0]=-1/(Set_C[2]-Set_C[0])*(tmp_v-Set_C[0])+1;
  else;
  if (tmp_v>=Set_C[1]&&tmp_v<Set_C[3])
  A_v[1]=1/(Set_C[3]-Set_C[1])*(tmp_v-Set_C[3])+1;
  else if (tmp_v>=Set_C[3]&&tmp_v<Set_C[5])
  A_v[1]=-1/(Set_C[5]-Set_C[3])*(tmp_v-Set_C[3])+1;
```

```
  else;
  if (tmp_v>=Set_C[4]&&tmp_v<Set_C[6])
  A_v[2]=1/(Set_C[6]-Set_C[4])*(tmp_v-Set_C[6])+1;
  else if (tmp_v>=Set_C[6])
  A_v[2]=1;
  else;
////// The End of The Third Table //////

////// The Fouth Table: dv //////
  if (tmp_dv<Set_D[0])
  A_dv[0]=1;
  else if (tmp_dv>=Set_D[0]&&tmp_dv<=Set_D[2])
  A_dv[0]=-1/(Set_D[2]-Set_D[0])*(tmp_dv-Set_D[0])+1;
  else;
  if (tmp_dv>=Set_D[1]&&tmp_dv<Set_D[3])
  A_dv[1]=1/(Set_D[3]-Set_D[1])*(tmp_dv-Set_D[3])+1;
  else if (tmp_dv>=Set_D[3]&&tmp_dv<Set_D[5])
  A_dv[1]=-1/(Set_D[5]-Set_D[3])*(tmp_dv-Set_D[3])+1;
  else;
  if (tmp_dv>=Set_D[4]&&tmp_dv<Set_D[6])
  A_dv[2]=1/(Set_D[6]-Set_D[4])*(tmp_dv-Set_D[6])+1;
  else if (tmp_dv>=Set_D[6])
  A_dv[2]=1;
  else;
////// The End of The Fouth Table //////

////// The Fifth Table: ap //////
  if (tmp_ap>=Set_E[0]&&tmp_ap<=Set_E[2])
  A_ap[0]=-1/(Set_E[2]-Set_E[0])*(tmp_ap-Set_E[0])+1;
  else;
  if (tmp_ap>=Set_E[1]&&tmp_ap<Set_E[3])
  A_ap[1]=1/(Set_E[3]-Set_E[1])*(tmp_ap-Set_E[3])+1;
  else if (tmp_ap>=Set_E[3]&&tmp_ap<Set_E[5])
  A_ap[1]=-1/(Set_E[5]-Set_E[3])*(tmp_ap-Set_E[3])+1;
  else;
  if (tmp_ap>=Set_E[4]&&tmp_ap<Set_E[6])
  A_ap[2]=1/(Set_E[6]-Set_E[4])*(tmp_ap-Set_E[6])+1;
  else if (tmp_ap>=Set_E[6])
  A_ap[2]=1;
  else;
////// The End of The Fifth Table //////
```

```
////// The Sixth Table: av //////
  if (tmp_av>=Set_F[0]&&tmp_av<=Set_F[2])
  A_av[0]=-1/(Set_F[2]-Set_F[0])*(tmp_av-Set_F[0])+1;
  else;
  if (tmp_av>=Set_F[1]&&tmp_av<Set_F[3])
  A_av[1]=1/(Set_F[3]-Set_F[1])*(tmp_av-Set_F[3])+1;
  else if (tmp_av>=Set_F[3]&&tmp_av<Set_F[5])
  A_av[1]=-1/(Set_F[5]-Set_F[3])*(tmp_av-Set_F[3])+1;
  else;
  if (tmp_av>=Set_F[4]&&tmp_av<Set_F[6])
  A_av[2]=1/(Set_F[6]-Set_F[4])*(tmp_av-Set_F[6])+1;
  else if (tmp_av>=Set_F[6])
  A_av[2]=1;
  else;
////// The End of The Sixth Table //////


////// Simplified Fuzzy Inference ///////
  sumdu=0;
  for (j=0;j<4;j++)
  {
    sumdu1[j]=0;
    sumdu2[j]=0;
    sumdw1[j]=0;
    sumdw2[j]=0;
    switch (j)
    {
      case 0:
        for (m=0;m<3;m++)
        {
          tmp_A1[m]=A_p[m];
          tmp_du[m]=du_p[m];
          tmp_A2[m]=A_ap[m];
          tmp_dw[m]=dw_p[m];
        }
        break;
      case 1:
        for (m=0;m<3;m++)
        {
          tmp_A1[m]=A_dp[m];
          tmp_du[m]=du_dp[m];
          tmp_A2[m]=A_ap[m];
```

```
          tmp_dw[m]=dw_dp[m];
        }
        break;
      case 2:
        for (m=0;m<3;m++)
        {
          tmp_A1[m]=A_v[m];
          tmp_du[m]=du_v[m];
          tmp_A2[m]=A_av[m];
          tmp_dw[m]=dw_v[m];
        }
        break;
      case 3:
        for (m=0;m<3;m++)
        {
          tmp_A1[m]=A_dv[m];
          tmp_du[m]=du_dv[m];
          tmp_A2[m]=A_av[m];
          tmp_dw[m]=dw_dv[m];
        }
        break;
    }

    for (i=0;i<3;i++)
    {
      sumdu1[j]=sumdu1[j]+tmp_A1[i]*tmp_du[i];
      sumdu2[j]=sumdu2[j]+tmp_A1[i];

      sumdw1[j]=sumdw1[j]+tmp_A2[i]*tmp_dw[i];
      sumdw2[j]=sumdw2[j]+tmp_A2[i];
    }
    du0[j]=sumdu1[j]/sumdu2[j];
    dw0[j]=sumdw1[j]/sumdw2[j];
    wD[j]=wB[j]+B[j]*dw0[j];
    sumdu=sumdu+wD[j]*du0[j];
  }
  du=sumdu;
  u[0]=lu[0]/1.0+du;
}

#define MDL_UPDATE /* Change to #undef to remove function */
```

```
#if defined(MDL_UPDATE)
/*Function: mdlUpdate ==========================================
 * Abstract:
 *  This function is called once for every major integration time step.
 *  Discrete states are typically updated here, but this function is useful
 *  for performing any tasks that should only take place once per
 *  integration step.
 */
static void mdlUpdate(SimStruct *S, int_T tid)
{
}
#endif /* MDL_UPDATE */

#define MDL_DERIVATIVES /* Change to #undef to remove function */
#if defined(MDL_DERIVATIVES)
/*Function: mdlDerivatives =====================================
 * Abstract:
 *  In this function, you compute the S-function block's derivatives.
 *  The derivatives are placed in the derivative vector, ssGetdX(S).
 */
static void mdlDerivatives(SimStruct *S)
{
}
#endif /* MDL_DERIVATIVES */

/*Function: mdlTerminate =======================================
 * Abstract:
 *  In this function, you should perform any actions that are necessary
 *  at the termination of a simulation. For example, if memory was
 *  allocated in mdlStart, this is the place to free it.
 */
static void mdlTerminate(SimStruct *S)
{
}

/*=========================================================*
 * See sfuntmpl_doc.c for the optional S-function methods *
 *=========================================================*/
/*=============================*
 * Required S-function trailer *
 *=============================*/
#ifdef MATLAB_MEX_FILE  /* Is this file being compiled as a MEX-file? */
```

```
#include "simulink.c"   /* MEX-file interface mechanism */
#else
#include "cg_sfun.h"   /* Code generation registration function */
#endif
```

4) Program running result (Fixed ID)

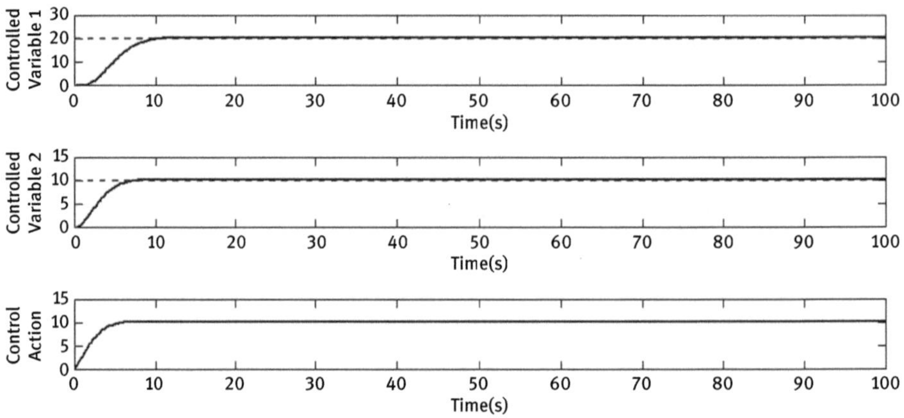

Figure 5.20: Simulative results of the SIRMs-based fuzzy control with fixed ID.

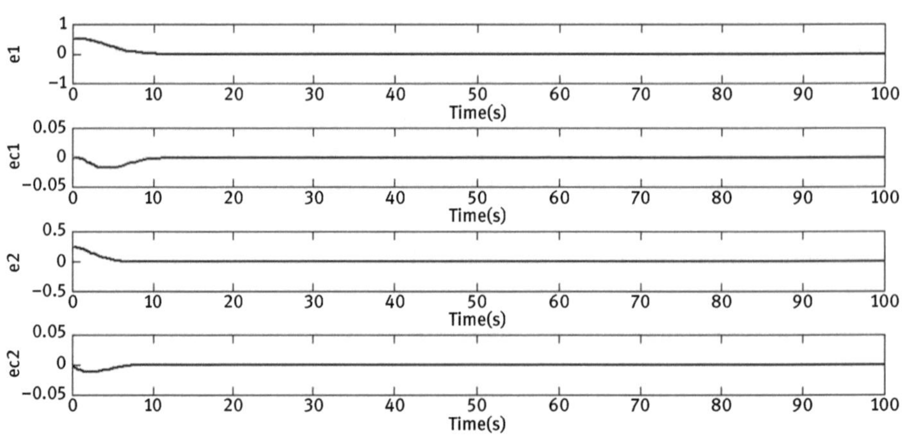

Figure 5.21: Errors and change of errors in the SIRMs-based fuzzy control with fixed ID.

5.5.2 SIRMs-based dynamically connected fuzzy inference control (DID, Version 1)

1) Simulink diagram of the SIRMs-based dynamically connected fuzzy inference control (DID, Version 1)

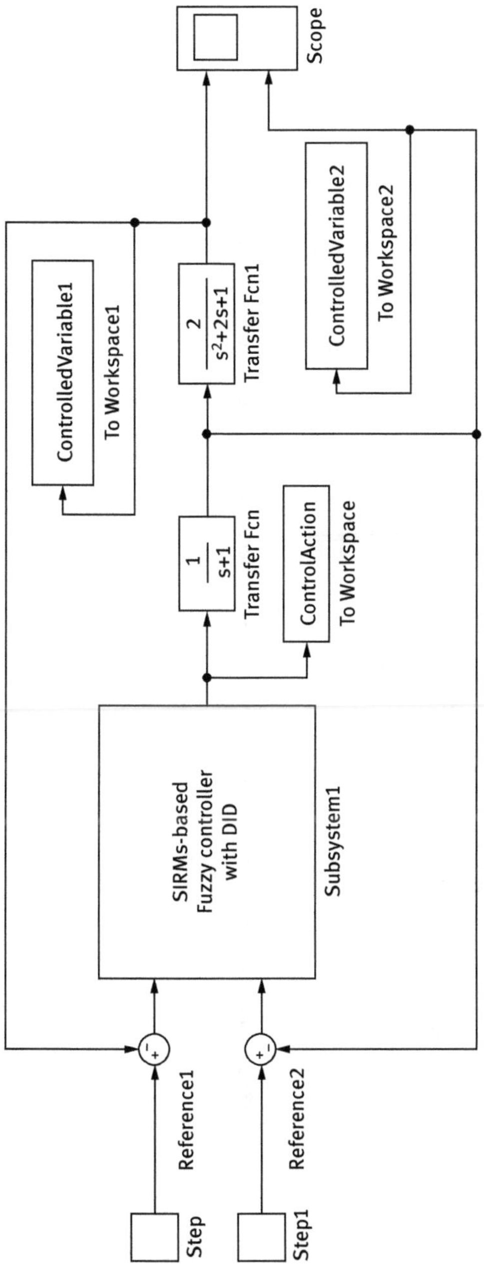

Figure 5.22: Errors and change of errors in the SIRMs-based fuzzy control with fixed ID.

2) Simulink diagram of the SIRMs-based fuzzy controller subsystem (DID, Version 1)

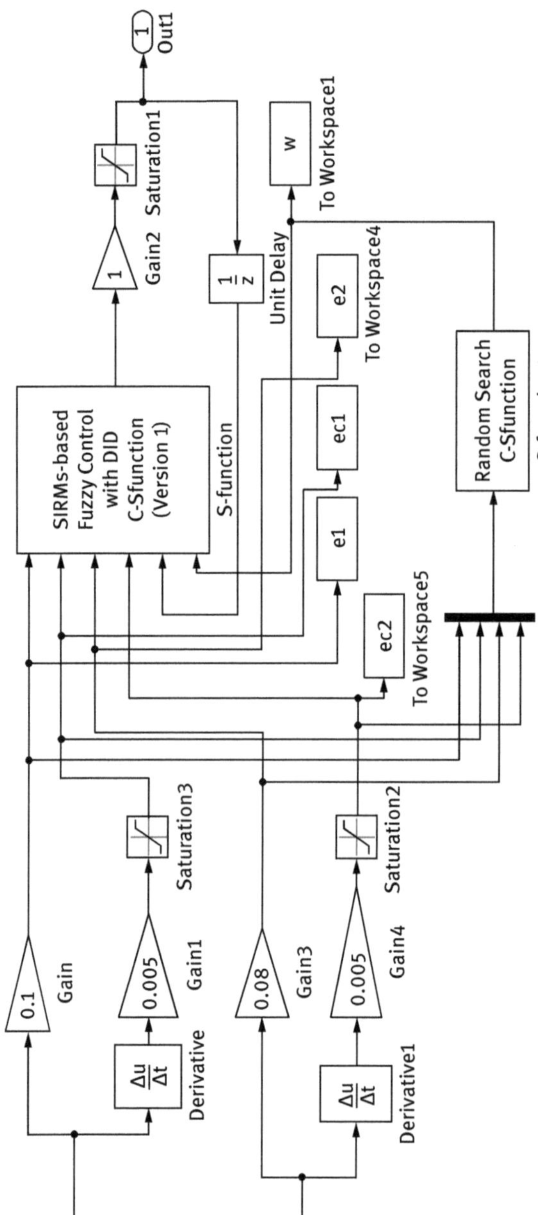

Figure 5.23: Simulink diagram of the SIRMs-based dynamically connected fuzzy inference control (DID, Version 1).

3) C-Sfunction for the SIRMs-based dynamically connected fuzzy inference control (DID)

```
    /*
  * SIRMs-based dynamically connected fuzzy inference control with DID
(Version 1)
  */

 /*
  * You must specify the S_FUNCTION_NAME as the name of your S-function
  * (i.e. replace sfuntmpl_basic with the name of your S-function).
  */

  #define S_FUNCTION_NAME sfSIRMFuzzyConDIDv1
  #define S_FUNCTION_LEVEL 2

 /*
  * Need to include simstruc.h for the definition of the SimStruct and
  * its associated macro definitions.
  */
  #include "simstruc.h"
  #include "math.h"

  #define sampletime(S) ssGetSFcnParam(S, 0)
 /* Error handling
  * --------------
  *
  * You should use the following technique to report errors encountered within
  * an S-function:
  *
  *   ssSetErrorStatus(S,"Error encountered due to ...");
  *   return;
  *
  * Note that the 2nd argument to ssSetErrorStatus must be persistent memory.
  * It cannot be a local variable. For example the following will cause
  * unpredictable errors:
  *
  *   mdlOutputs()
  *   {
  *     char msg[256];      {ILLEGAL: to fix use "static char msg[256];"}
  *     sprintf(msg,"Error due to %s", string);
  *     ssSetErrorStatus(S,msg);
```

```
 *     return;
 *   }
 *
 * See matlabroot/simulink/src/sfuntmpl_doc.c for more details.
 */

/*=====================*
 * S-function methods *
 *=====================*/
/* Function: mdlInitializeSizes ========================================
 * Abstract:
 *  The sizes information is used by Simulink to determine the S-function
 *  block's characteristics (number of inputs, outputs, states, etc.).
 */
 static int_T count;

 static void mdlInitializeSizes(SimStruct *S)
 {
   /* See sfuntmpl_doc.c for more details on the macros below */

   ssSetNumSFcnParams(S, 1); /* Number of expected parameters */
   if (ssGetNumSFcnParams(S) != ssGetSFcnParamsCount(S)) {
     /* Return if number of expected != number of actual parameters */
     return;
   }

   ssSetNumContStates(S, 0);
   ssSetNumDiscStates(S, 0);

   if (!ssSetNumInputPorts(S, 6)) return;  //Specify the number of input
ports that a block has.
   ssSetInputPortWidth(S, 0, 1);      //Specify the width of an input port.
   ssSetInputPortWidth(S, 1, 1);      //Specify the width of an input port.
   ssSetInputPortWidth(S, 2, 1);      //Specify the width of an input port.
   ssSetInputPortWidth(S, 3, 1);      //Specify the width of an input port.
   ssSetInputPortWidth(S, 4, 1);      //Specify the width of an input port.
   ssSetInputPortWidth(S, 5, 8);      //Specify the width of an input port.
   ssSetInputPortRequiredContiguous(S, 0, true); /*direct input signal
access*/
   ssSetInputPortRequiredContiguous(S, 1, true); /*direct input signal
access*/
```

```
    ssSetInputPortRequiredContiguous(S, 2, true); /*direct input signal
access*/
    ssSetInputPortRequiredContiguous(S, 3, true); /*direct input signal
access*/
    ssSetInputPortRequiredContiguous(S, 4, true); /*direct input signal
access*/
    ssSetInputPortRequiredContiguous(S, 5, true); /*direct input signal
access*/
    /*
     * Set direct feedthrough flag (1=yes, 0=no).
     * A port has direct feedthrough if the input is used in either
     * the mdlOutputs or mdlGetTimeOfNextVarHit functions.
     * See matlabroot/simulink/src/sfuntmpl_directfeed.txt.
     */
    ssSetInputPortDirectFeedThrough(S, 0, 1); //Specify the direct feed-
through status of a block's ports.
    ssSetInputPortDirectFeedThrough(S, 1, 1); //Specify the direct feed-
through status of a block's ports.
    ssSetInputPortDirectFeedThrough(S, 2, 1); //Specify the direct feed-
through status of a block's ports.
    ssSetInputPortDirectFeedThrough(S, 3, 1); //Specify the direct feed-
through status of a block's ports.
    ssSetInputPortDirectFeedThrough(S, 4, 1); //Specify the direct feed-
through status of a block's ports.
    ssSetInputPortDirectFeedThrough(S, 5, 1); //Specify the direct feed-
through status of a block's ports.

    if (!ssSetNumOutputPorts(S, 1)) return;
    ssSetOutputPortWidth(S, 0, 1);

    ssSetNumSampleTimes(S, 1);      //Specify the number of sample times that
an S-Function block has.
    ssSetNumRWork(S, 0);
    ssSetNumIWork(S, 0);
    ssSetNumPWork(S, 0);
    ssSetNumModes(S, 0);
    ssSetNumNonsampledZCs(S, 0);

    ssSetOptions(S, 0);

  }

  /* Function: mdlInitializeSampleTimes =================================
   * Abstract:
```

```
 *  This function is used to specify the sample time(s) for your
 *  S-function. You must register the same number of sample times as
 *  specified in ssSetNumSampleTimes.
 */
static void mdlInitializeSampleTimes(SimStruct *S)
{
  const real_T  *stime=mxGetPr(sampletime(S));
  ssSetSampleTime(S, 0, stime[0]);
  ssSetOffsetTime(S, 0, 0.0);
}

#define MDL_INITIALIZE_CONDITIONS /* Change to #undef to remove function */
#if defined(MDL_INITIALIZE_CONDITIONS)
/* Function: mdlInitializeConditions ==================================
 * Abstract:
 *  In this function, you should initialize the continuous and discrete
 *  states for your S-function block. The initial states are placed
 *  in the state vector, ssGetContStates(S) or ssGetRealDiscStates(S).
 *  You can also perform any other initialization activities that your
 *  S-function may require. Note, this routine will be called at the
 *  start of simulation and if it is present in an enabled subsystem
 *  configured to reset states, it will be call when the enabled subsystem
 *  restarts execution to reset the states.
 */
static void mdlInitializeConditions(SimStruct *S)
{
}
#endif /* MDL_INITIALIZE_CONDITIONS */

#define MDL_START /* Change to #undef to remove function */
#if defined(MDL_START)
/* Function: mdlStart ================================================
 * Abstract:
 *  This function is called once at start of model execution. If you
 *  have states that should be initialized once, this is the place
 *  to do it.
 */
static void mdlStart(SimStruct *S)
{
  count=0;
}
#endif /* MDL_START */
```

```
/* Function: mdlOutputs =========================================
 * Abstract:
 *  In this function, you compute the outputs of your S-function
 *  block. Generally outputs are placed in the output vector, ssGetY(S).
 */

 static void mdlOutputs(SimStruct *S, int_T tid)
 {
   const real_T  *p = ssGetInputPortRealSignal(S,0);
   const real_T  *dp = ssGetInputPortRealSignal(S,1);
   const real_T  *v = ssGetInputPortRealSignal(S,2);
   const real_T  *dv = ssGetInputPortRealSignal(S,3);
   const real_T  *lu = ssGetInputPortRealSignal(S,4);
   const real_T  *wi = ssGetInputPortRealSignal(S,5);
   const real_T  *stime = mxGetPr(sampletime(S));
   real_T     *u = (real_T *)ssGetOutputPortSignal(S,0);
   real_T Set_A[7] = {-1,-1,0,0,0,1,1};
   real_T Set_B[7] = {-1,-1,0,0,0,1,1};
   real_T Set_C[7] = {-1,-1,0,0,0,1,1};
   real_T Set_D[7] = {-1,-1,0,0,0,1,1};
   real_T Set_E[7] = {0,0,0.5,0.5,0.5,1,1};
   real_T Set_F[7] = {0,0,0.5,0.5,0.5,1,1};
   real_T A_p[3]  = {0,0,0};
   real_T A_dp[3] = {0,0,0};
   real_T A_v[3]  = {0,0,0};
   real_T A_dv[3] = {0,0,0};
   real_T A_ap[3] = {0,0,0};
   real_T A_av[3] = {0,0,0};
   real_T du_p[3] = {-1,0,1};
   real_T du_dp[3] = {-1,0,1};
   real_T du_v[3] = {-1,0,1};
   real_T du_dv[3] = {-1,0,1};
   real_T dw_p[3] = {0,0.5,1};
   real_T dw_dp[3] = {0,0.5,1};
   real_T dw_v[3] = {0,0.5,1};
   real_T dw_dv[3] = {0,0.5,1};
   real_T wB[4]   = {0,0,0,0};
   real_T B[4]   = {1,1,1,1};
   real_T du0[4],dw0[4],du,wD[4];
   real_T tmp_p,tmp_dp,tmp_v,tmp_dv,tmp_ap,tmp_av,sumdu;
   real_T sumdu1[4],sumdu2[4],sumdw1[4],sumdw2[4];
   real_T tmp_A1[3],tmp_du[3],tmp_A2[3],tmp_dw[3];
```

```
  int_T i,j,m;

////// The End of The Forth Table //////
  tmp_p = p[0];
  tmp_dp = dp[0];
  tmp_v = v[0];
  tmp_dv = dv[0];
  tmp_ap = fabs(p[0]);
  tmp_av = fabs(v[0]);
  for (i=0;i<4;i++)
  {
    wB[i]=wi[i];
    B[i]=wi[4+i];
  }

////// The First Table: p //////
  if (tmp_p<Set_A[0])
  A_p[0]=1;
  else if (tmp_p>=Set_A[0]&&tmp_p<=Set_A[2])
  A_p[0]=-1/(Set_A[2]-Set_A[0])*(tmp_p-Set_A[0])+1;
  else;
  if (tmp_p>=Set_A[1]&&tmp_p<Set_A[3])
  A_p[1]=1/(Set_A[3]-Set_A[1])*(tmp_p-Set_A[3])+1;
  else if (tmp_p>=Set_A[3]&&tmp_p<Set_A[5])
  A_p[1]=-1/(Set_A[5]-Set_A[3])*(tmp_p-Set_A[3])+1;
  else;
  if (tmp_p>=Set_A[4]&&tmp_p<Set_A[6])
  A_p[2]=1/(Set_A[6]-Set_A[4])*(tmp_p-Set_A[6])+1;
  else if (tmp_p>=Set_A[6])
  A_p[2]=1;
  else;
////// The End of The First Table //////

////// The Second Table: dp //////
  if (tmp_dp<Set_B[0])
  A_dp[0]=1;
  else if (tmp_dp>=Set_B[0]&&tmp_dp<=Set_B[2])
  A_dp[0]=-1/(Set_B[2]-Set_B[0])*(tmp_dp-Set_B[0])+1;
  else;
  if (tmp_dp>=Set_B[1]&&tmp_dp<Set_B[3])
  A_dp[1]=1/(Set_B[3]-Set_B[1])*(tmp_dp-Set_B[3])+1;
  else if (tmp_dp>=Set_B[3]&&tmp_dp<Set_B[5])
```

```
  A_dp[1]=-1/(Set_B[5]-Set_B[3])*(tmp_dp-Set_B[3])+1;
  else;
  if (tmp_dp>=Set_B[4]&&tmp_dp<Set_B[6])
  A_dp[2]=1/(Set_B[6]-Set_B[4])*(tmp_dp-Set_B[6])+1;
  else if (tmp_dp>=Set_B[6])
  A_dp[2]=1;
  else;
////// The End of The Second Table //////

////// The Third Table: v //////
  if (tmp_v<Set_C[0])
  A_v[0]=1;
  else if (tmp_v>=Set_C[0]&&tmp_v<=Set_C[2])
  A_v[0]=-1/(Set_C[2]-Set_C[0])*(tmp_v-Set_C[0])+1;
  else;
  if (tmp_v>=Set_C[1]&&tmp_v<Set_C[3])
  A_v[1]=1/(Set_C[3]-Set_C[1])*(tmp_v-Set_C[3])+1;
  else if (tmp_v>=Set_C[3]&&tmp_v<Set_C[5])
  A_v[1]=-1/(Set_C[5]-Set_C[3])*(tmp_v-Set_C[3])+1;
  else;
  if (tmp_v>=Set_C[4]&&tmp_v<Set_C[6])
  A_v[2]=1/(Set_C[6]-Set_C[4])*(tmp_v-Set_C[6])+1;
  else if (tmp_v>=Set_C[6])
  A_v[2]=1;
  else;
////// The End of The Third Table //////

////// The Fouth Table: dv //////
  if (tmp_dv<Set_D[0])
  A_dv[0]=1;
  else if (tmp_dv>=Set_D[0]&&tmp_dv<=Set_D[2])
  A_dv[0]=-1/(Set_D[2]-Set_D[0])*(tmp_dv-Set_D[0])+1;
  else;
  if (tmp_dv>=Set_D[1]&&tmp_dv<Set_D[3])
  A_dv[1]=1/(Set_D[3]-Set_D[1])*(tmp_dv-Set_D[3])+1;
  else if (tmp_dv>=Set_D[3]&&tmp_dv<Set_D[5])
  A_dv[1]=-1/(Set_D[5]-Set_D[3])*(tmp_dv-Set_D[3])+1;
  else;
  if (tmp_dv>=Set_D[4]&&tmp_dv<Set_D[6])
  A_dv[2]=1/(Set_D[6]-Set_D[4])*(tmp_dv-Set_D[6])+1;
  else if (tmp_dv>=Set_D[6])
  A_dv[2]=1;
```

```
  else;
////// The End of The Fouth Table //////

////// The Fifth Table: ap //////
  if (tmp_ap>=Set_E[0]&&tmp_ap<=Set_E[2])
  A_ap[0]=-1/(Set_E[2]-Set_E[0])*(tmp_ap-Set_E[0])+1;
  else;
  if (tmp_ap>=Set_E[1]&&tmp_ap<Set_E[3])
  A_ap[1]=1/(Set_E[3]-Set_E[1])*(tmp_ap-Set_E[3])+1;
  else if (tmp_ap>=Set_E[3]&&tmp_ap<Set_E[5])
  A_ap[1]=-1/(Set_E[5]-Set_E[3])*(tmp_ap-Set_E[3])+1;
  else;
  if (tmp_ap>=Set_E[4]&&tmp_ap<Set_E[6])
  A_ap[2]=1/(Set_E[6]-Set_E[4])*(tmp_ap-Set_E[6])+1;
  else if (tmp_ap>=Set_E[6])
  A_ap[2]=1;
  else;
////// The End of The Fifth Table //////

////// The Sixth Table: av //////
  if (tmp_av>=Set_F[0]&&tmp_av<=Set_F[2])
  A_av[0]=-1/(Set_F[2]-Set_F[0])*(tmp_av-Set_F[0])+1;
  else;
  if (tmp_av>=Set_F[1]&&tmp_av<Set_F[3])
  A_av[1]=1/(Set_F[3]-Set_F[1])*(tmp_av-Set_F[3])+1;
  else if (tmp_av>=Set_F[3]&&tmp_av<Set_F[5])
  A_av[1]=-1/(Set_F[5]-Set_F[3])*(tmp_av-Set_F[3])+1;
  else;
  if (tmp_av>=Set_F[4]&&tmp_av<Set_F[6])
  A_av[2]=1/(Set_F[6]-Set_F[4])*(tmp_av-Set_F[6])+1;
  else if (tmp_av>=Set_F[6])
  A_av[2]=1;
  else;
////// The End of The Sixth Table //////

////// Simplified Fuzzy Inference ///////
  sumdu=0;
  for (j=0;j<4;j++)
  {
    sumdu1[j]=0;
    sumdu2[j]=0;
    sumdw1[j]=0;
```

```
sumdw2[j]=0;
switch (j)
{
  case 0:
    for (m=0;m<3;m++)
    {
      tmp_A1[m]=A_p[m];
      tmp_du[m]=du_p[m];
      tmp_A2[m]=A_ap[m];
      tmp_dw[m]=dw_p[m];
    }
    break;
  case 1:
    for (m=0;m<3;m++)
    {
      tmp_A1[m]=A_dp[m];
      tmp_du[m]=du_dp[m];
      tmp_A2[m]=A_ap[m];
      tmp_dw[m]=dw_dp[m];
    }
    break;
  case 2:
    for (m=0;m<3;m++)
    {
      tmp_A1[m]=A_v[m];
      tmp_du[m]=du_v[m];
      tmp_A2[m]=A_av[m];
      tmp_dw[m]=dw_v[m];
    }
    break;
  case 3:
    for (m=0;m<3;m++)
    {
      tmp_A1[m]=A_dv[m];
      tmp_du[m]=du_dv[m];
      tmp_A2[m]=A_av[m];
      tmp_dw[m]=dw_dv[m];
    }
    break;
}

for (i=0;i<3;i++)
```

```
      {
        sumdu1[j]=sumdu1[j]+tmp_A1[i]*tmp_du[i];
        sumdu2[j]=sumdu2[j]+tmp_A1[i];

        sumdw1[j]=sumdw1[j]+tmp_A2[i]*tmp_dw[i];
        sumdw2[j]=sumdw2[j]+tmp_A2[i];
      }
      du0[j]=sumdu1[j]/sumdu2[j];
      dw0[j]=sumdw1[j]/sumdw2[j];
      wD[j]=wB[j]+B[j]*dw0[j];
      sumdu=sumdu+wD[j]*du0[j];
    }

    du=sumdu;
    u[0]=lu[0]/1.5+du;
  }

  #define MDL_UPDATE /* Change to #undef to remove function */
  #if defined(MDL_UPDATE)
 /* Function: mdlUpdate ===============================================
  * Abstract:
  *  This function is called once for every major integration time step.
  *  Discrete states are typically updated here, but this function is useful
  *  for performing any tasks that should only take place once per
  *  integration step.
  */
  static void mdlUpdate(SimStruct *S, int_T tid)
  {
  }
  #endif /* MDL_UPDATE */

 /* Function: mdlTerminate ============================================
  * Abstract:
  *  In this function, you should perform any actions that are necessary
  *  at the termination of a simulation. For example, if memory was
  *  allocated in mdlStart, this is the place to free it.
  */
  static void mdlTerminate(SimStruct *S)
  {
  }

 /*=========================================================*
```

```
 * See sfuntmpl_doc.c for the optional S-function methods *
 *=========================================================*/

/*===============================*
 * Required S-function trailer *
 *===============================*/

 #ifdef MATLAB_MEX_FILE  /* Is this file being compiled as a MEX-file? */
 #include "simulink.c"   /* MEX-file interface mechanism */
 #else
 #include "cg_sfun.h"    /* Code generation registration function */
 #endif
/*
 * Random Search Algorithm
 */

/*
 * You must specify the S_FUNCTION_NAME as the name of your S-function
 * (i.e. replace sfuntmpl_basic with the name of your S-function).
 */

 #define S_FUNCTION_NAME sfRandomSearch
 #define S_FUNCTION_LEVEL 2
/*
 * Need to include simstruc.h for the definition of the SimStruct and
 * its associated macro definitions.
 */
 #include "simstruc.h"
 #include "math.h"
 #include "sfRandomSearch.h"

 #define sampletime(S) ssGetSFcnParam(S, 0)

/* Error handling
 * --------------
 *
 * You should use the following technique to report errors encountered within
 * an S-function:
 *
 *   ssSetErrorStatus(S,"Error encountered due to ...");
 *   return;
 *
```

```
 * Note that the 2nd argument to ssSetErrorStatus must be persistent memory.
 * It cannot be a local variable. For example the following will cause
 * unpredictable errors:
 *
 *  mdlOutputs()
 *  {
 *    char msg[256];      {ILLEGAL: to fix use "static char msg[256];"}
 *    sprintf(msg,"Error due to %s", string);
 *    ssSetErrorStatus(S,msg);
 *    return;
 *  }
 *
 * See matlabroot/simulink/src/sfuntmpl_doc.c for more details.
 */

/*====================*
 * S-function methods *
 *====================*/

/* Function: mdlInitializeSizes ===============================================
 * Abstract:
 *  The sizes information is used by Simulink to determine the S-function
 *  block's characteristics (number of inputs, outputs, states, etc.).
 */
 static int_T count;
 static int_T nostep;
 static int_T failure;
 static int_T success;

 static void mdlInitializeSizes(SimStruct *S)
 {
   /* See sfuntmpl_doc.c for more details on the macros below */
    ssSetNumSFcnParams(S, 1); /* Number of expected parameters */
   if (ssGetNumSFcnParams(S) != ssGetSFcnParamsCount(S)) {
     /* Return if number of expected != number of actual parameters */
     return;
   }

   ssSetNumContStates(S, 0);
   ssSetNumDiscStates(S, 1);

   if (!ssSetNumInputPorts(S, 1)) return;  //Specify the number of input
ports that a block has.
```

```
    ssSetInputPortWidth(S, 0, 4);       //Specify the width of an input port.
    ssSetInputPortRequiredContiguous(S, 0, true); /*direct input signal
access*/
    /*
     * Set direct feedthrough flag (1=yes, 0=no).
     * A port has direct feedthrough if the input is used in either
     * the mdlOutputs or mdlGetTimeOfNextVarHit functions.
     * See matlabroot/simulink/src/sfuntmpl_directfeed.txt.
     */
    ssSetInputPortDirectFeedThrough(S, 0, 1); //Specify the direct feed-
through status of a block's ports.
    if (!ssSetNumOutputPorts(S, 1)) return;
    ssSetOutputPortWidth(S, 0, 8);

    ssSetNumSampleTimes(S, 1);       //Specify the number of sample times that
an S-Function block has.
    ssSetNumRWork(S, 0);
    ssSetNumIWork(S, 0);
    ssSetNumPWork(S, 0);
    ssSetNumModes(S, 0);
    ssSetNumNonsampledZCs(S, 0);
    ssSetOptions(S, 0);
  }

 /* Function: mdlInitializeSampleTimes =================================
  * Abstract:
  *  This function is used to specify the sample time(s) for your
  *  S-function. You must register the same number of sample times as
  *  specified in ssSetNumSampleTimes.
  */
  static void mdlInitializeSampleTimes(SimStruct *S)
  {
    const real_T  *stime=mxGetPr(sampletime(S));
    ssSetSampleTime(S, 0, stime[0]);
    ssSetOffsetTime(S, 0, 0.0);
  }

  #define MDL_INITIALIZE_CONDITIONS /* Change to #undef to remove function
  */
  #if defined(MDL_INITIALIZE_CONDITIONS)
 /* Function: mdlInitializeConditions ==================================
  * Abstract:
```

```
 *  In this function, you should initialize the continuous and discrete
 *  states for your S-function block. The initial states are placed
 *  in the state vector, ssGetContStates(S) or ssGetRealDiscStates(S).
 *  You can also perform any other initialization activities that your
 *  S-function may require. Note, this routine will be called at the
 *  start of simulation and if it is present in an enabled subsystem
 *  configured to reset states, it will be call when the enabled subsystem
 *  restarts execution to reset the states.
 */
 static void mdlInitializeConditions(SimStruct *S)
 {
 }
 #endif /* MDL_INITIALIZE_CONDITIONS */

 #define MDL_START /* Change to #undef to remove function */
 #if defined(MDL_START)
/* Function: mdlStart ===============================================
 * Abstract:
 *  This function is called once at start of model execution. If you
 *  have states that should be initialized once, this is the place
 *  to do it.
 */
 static void mdlStart(SimStruct *S)
 {
   count=0;
   nostep=0;
   failure=0;
   success=0;
 }
 #endif /* MDL_START */

/* Function: mdlOutputs =============================================
 * Abstract:
 *  In this function, you compute the outputs of your S-function
 *  block. Generally outputs are placed in the output vector, ssGetY(S).
 */
 static void mdlOutputs(SimStruct *S, int_T tid)
 {
   real_T *xi    = ssGetInputPortRealSignal(S,0);
   real_T *stime = mxGetPr(sampletime(S));
   real_T *xd    = ssGetRealDiscStates(S);
   real_T *xo    = (real_T *)ssGetOutputPortSignal(S,0);
```

```
    static real_T theta[8],ltheta[8],xrv[8],L[3],miu[8],lmiu[8],lxrv[8];
    real_T step1;
   int_T i;
    if (count==0)
    {
      for(i=0;i<8;i++)
      {
        ltheta[i]=((double)(rand()%1000)/1000.0)*(100 -0) +0;
        lxrv[i]=((double)(rand()%1000)/1000)*0.1;
        lmiu[i]=0;
      }
    }
    else if (nostep<4)
    {
      if (nostep==0)
      {
        for (i=0;i<8;i++)
        {
          step1=0;
          label: xrv[i]=box_muller(lxrv[i],lmiu[i]);
          if  ((((ltheta[i]+xrv[i])<=100)  &&  ((ltheta[i]+xrv[i])>=0))  &&
(((ltheta[i]-xrv[i])<=100) && ((ltheta[i]-xrv[i])>=0)))
            continue;
          else
          {
            if (step1>100){xrv[i]=lxrv[i];continue;}
            step1=step1+1;
          goto label;
          }
        }
      }
      switch(nostep)
      {
        case 0:
          for (i=0;i<8;i++)
            xo[i]=ltheta[i];
          break;
        case 1:
          for (i=0;i<8;i++)
            xo[i]=ltheta[i]+xrv[i];
            L[1]=evaluate(xi,0);
          break;
```

```
  case 2:
    for (i=0;i<8;i++)
      xo[i]=ltheta[i]-xrv[i];
      L[0]=evaluate(xi,0);
    break;
  case 3:
    L[2]=evaluate(xi,0);
}
nostep=nostep+1;
if (nostep==4)
{
  nostep=0;
  if (L[0]<L[1])
  {
    for(i=0;i<8;i++)
    {
      theta[i]=ltheta[i]+xrv[i];
      miu[i]=0.4*xrv[i]+0.2*lmiu[i];
    }
  }
  else if((L[2]<L[1]) && (L[0]>=L[1]))
  {
    for(i=0;i<8;i++)
    {
      theta[i]=ltheta[i]-xrv[i];
      miu[i]=lmiu[i]-0.4*xrv[i];
    }
  }
  else
  {
    for(i=0;i<8;i++)
    {
      theta[i]=ltheta[i];
      miu[i]=0.5*lmiu[i];
    }
  }
  for (i=0;i<8;i++)
  {
    xo[i]=theta[i];
    ltheta[i]=theta[i];
    lmiu[i]=miu[i];
    lxrv[i]=xrv[i];
```

```
      }
    }
    for(i=0;i<3;i++)
      printf("L[%d]=%f,%d\n",i,L[i],count);
    for(i=0;i<8;i++)
    {
      printf("xo[%d]=%f,xrv[%d]=%f\n",i,xo[i],i,xrv[i]);
    }
  }
  count=count+1;
}

#define MDL_UPDATE /* Change to #undef to remove function */
#if defined(MDL_UPDATE)
/* Function: mdlUpdate ==========================================
 * Abstract:
 *  This function is called once for every major integration time step.
 *  Discrete states are typically updated here, but this function is useful
 *  for performing any tasks that should only take place once per
 *  integration step.
 */
static void mdlUpdate(SimStruct *S, int_T tid)
{
}
#endif /* MDL_UPDATE */

#define MDL_DERIVATIVES /* Change to #undef to remove function */
#if defined(MDL_DERIVATIVES)
/* Function: mdlDerivatives =======================================
 * Abstract:
 *  In this function, you compute the S-function block's derivatives.
 *  The derivatives are placed in the derivative vector, ssGetdX(S).
 */
static void mdlDerivatives(SimStruct *S)
{
}
#endif /* MDL_DERIVATIVES */

/* Function: mdlTerminate ========================================
 * Abstract:
 *  In this function, you should perform any actions that are necessary
 *  at the termination of a simulation. For example, if memory was
```

```
 *  allocated in mdlStart, this is the place to free it.
 */
 static void mdlTerminate(SimStruct *S)
 {
 }

/*============================================================*
 * See sfuntmpl_doc.c for the optional S-function methods *
 *============================================================*/

/*================================*
 * Required S-function trailer *
 *================================*/

 #ifdef MATLAB_MEX_FILE  /* Is this file being compiled as a MEX-file? */
 #include "simulink.c"    /* MEX-file interface mechanism */
 #else
 #include "cg_sfun.h"    /* Code generation registration function */
 #endif

 //////////////////////////sfRandomSearch.h///////////////
 #include "math.h"
 #include "time.h"

 double evaluate(double *,double);
 double box_muller(double, double);
 double ranf();

 double evaluate(double *y,double z)
 {
   double fit;
   fit = z+fabs(y[0])+fabs(y[1])+fabs(y[2])+fabs(y[3]);
   //fit = z+y[0]*y[0]+y[1]*y[1]+y[2]*y[2]+y[3]*y[3];

   return (fit);
 }

 double box_muller(double m, double s) /* normal random variate generator */
 {            /* mean m, standard deviation s */
   double x1, x2, w, y1;
   static double y2;
   static int use_last = 0;
   if (use_last)            /* use value from previous call */
   {
     y1 = y2;
     use_last = 0;
```

```
    }
    else
    {
      do {
        x1 = 2.0 * ranf() - 1.0;
        x2 = 2.0 * ranf() - 1.0;
        w = x1 * x1 + x2 * x2;
      } while ( w >= 1.0 );

      w = sqrt( (-2.0 * log( w ) ) / w );
      y1 = x1 * w;
      y2 = x2 * w;
      use_last = 1;
    }
    return( m + y1 * s );
  }
  double ranf()
  {
  int i;
  double x;
  //srand( (unsigned)time( NULL ) );
  i = rand();
  x = (double) i / (double) RAND_MAX;
  return(x);
  }
```

4) Program running result (DID, Version 1)

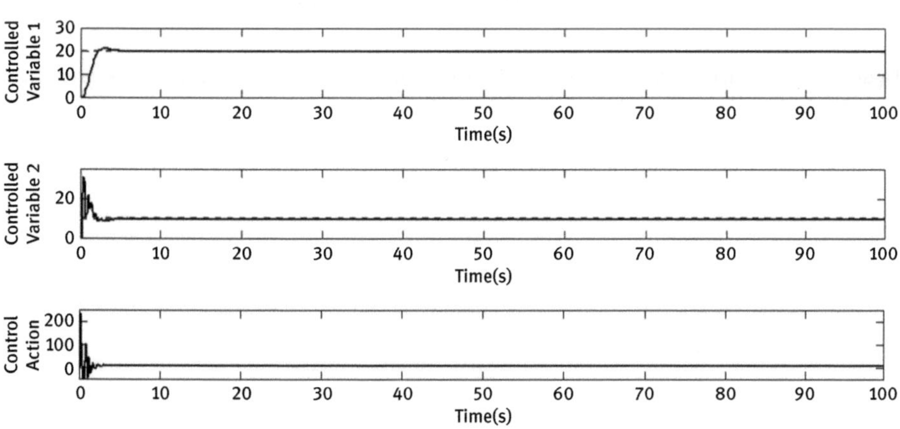

Figure 5.24: Simulink diagram of the SIRMs-based fuzzy controller subsystem (DID, Version 1).

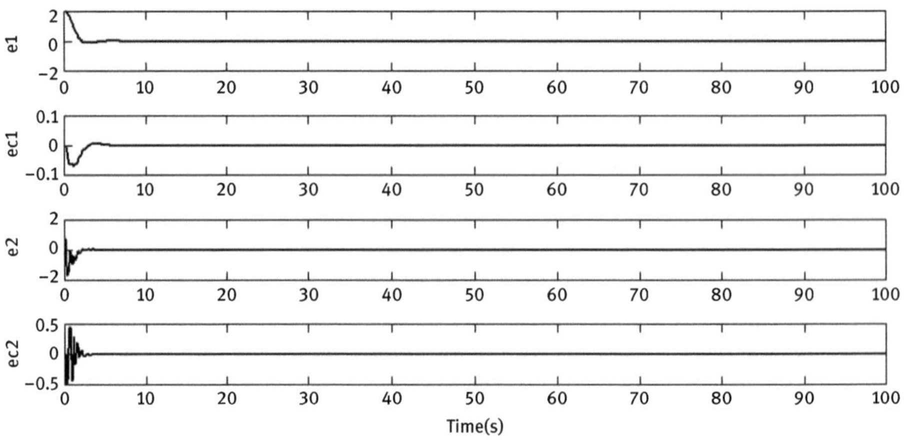

Figure 5.25: Errors and change of errors in the SIRMs-based fuzzy control with DID (Version 1).

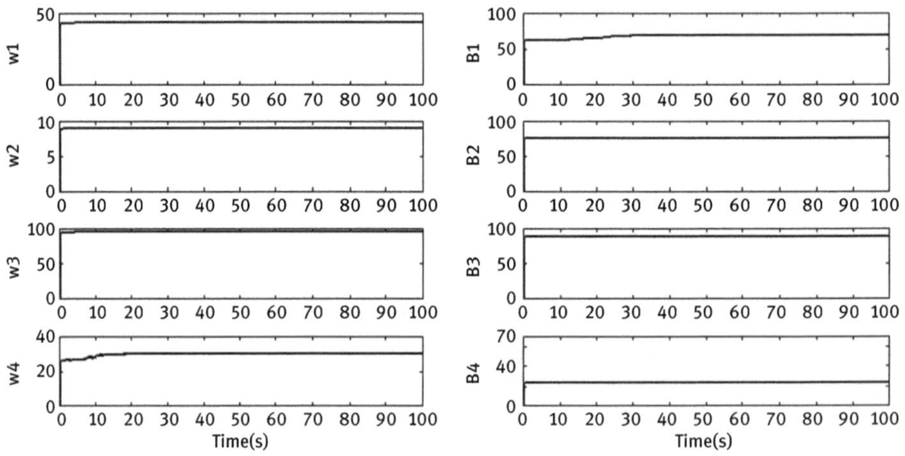

Figure 5.26: Optimized parameters: base value wi and breadth Bi, where i = 1, 2, 3, 4 (Version 1).

5.5.3 SIRMs-based dynamically connected fuzzy inference control (DID, Version 2)

1) Simulink diagram of the SIRMs-based dynamically connected fuzzy inference control (DID, Version 2)

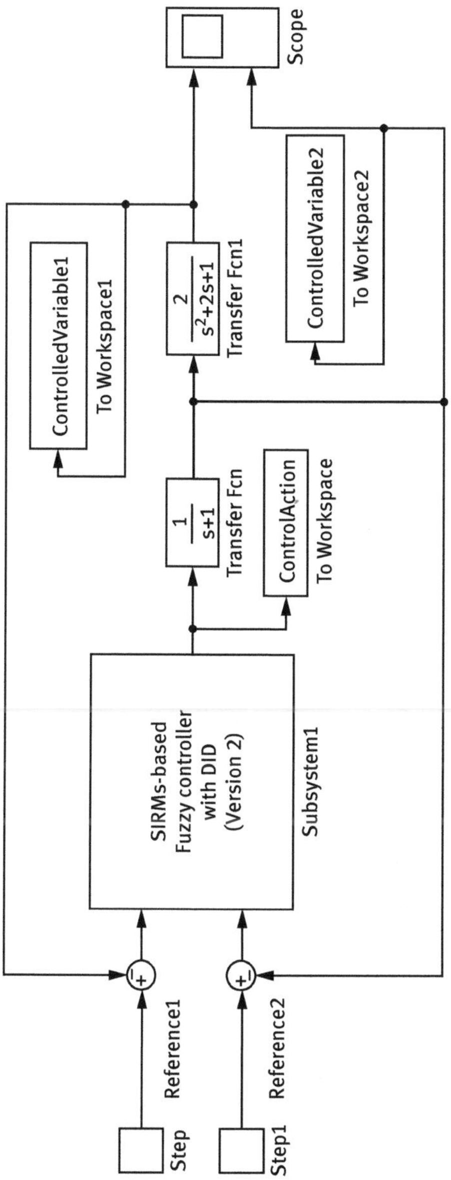

Figure 5.27: Simulink diagram of the SIRMs-based dynami-cally connected fuzzy inference control (DID, Version 2).

2) Simulink diagram of the SIRMs-based fuzzy controller subsystem (DID, Version 2)

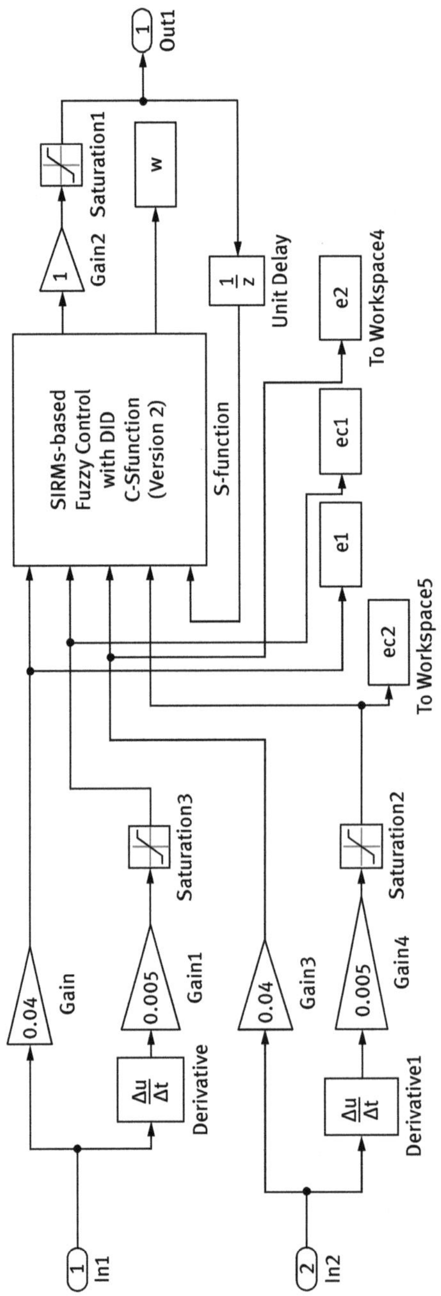

Figure 5.28: Simulink diagram of the SIRMs-based fuzzy controller subsystem (DID, Version 2).

3) C-Sfunction for the SIRMs-based dynamically connected fuzzy inference control (DID, Version 2)

```
/*
/* SIRMs-based Fuzzy Inference Control with DID (Version 2)
/*

 * You must specify the S_FUNCTION_NAME as the name of your S-function
 * (i.e. replace sfuntmpl_basic with the name of your S-function).
 */

 #define S_FUNCTION_NAME sfSIRMFuzzyControlDIDv2
 #define S_FUNCTION_LEVEL 2

/*
 * Need to include simstruc.h for the definition of the SimStruct and
 * its associated macro definitions.
 */
 #include "simstruc.h"
 #include "funRSde.h"

 #define sampletime(S) ssGetSFcnParam(S, 0)
/* Error handling
 * --------------
 *
 * You should use the following technique to report errors encountered within
 * an S-function:
 *
 *   ssSetErrorStatus(S,"Error encountered due to ...");
 *   return;
 *
 * Note that the 2nd argument to ssSetErrorStatus must be persistent memory.
 * It cannot be a local variable. For example the following will cause
 * unpredictable errors:
 *
 *   mdlOutputs()
 *   {
 *     char msg[256];      {ILLEGAL: to fix use "static char msg[256];"}
 *     sprintf(msg,"Error due to %s", string);
 *     ssSetErrorStatus(S,msg);
 *     return;
 *   }
 *
```

```
 * See matlabroot/simulink/src/sfuntmpl_doc.c for more details.
 */
/*====================*
 * S-function methods *
 *====================*/

/* Function: mdlInitializeSizes ===========================================
 * Abstract:
 *  The sizes information is used by Simulink to determine the S-function
 *  block's characteristics (number of inputs, outputs, states, etc.).
 */
 static int_T count;
 static int_T nostep;
 static int_T label;

 static void mdlInitializeSizes(SimStruct *S)
 {
   /* See sfuntmpl_doc.c for more details on the macros below */
   ssSetNumSFcnParams(S, 1); /* Number of expected parameters */
   if (ssGetNumSFcnParams(S) != ssGetSFcnParamsCount(S)) {
     /* Return if number of expected != number of actual parameters */
     return;
   }

   ssSetNumContStates(S, 0);
   ssSetNumDiscStates(S, 0);

   if (!ssSetNumInputPorts(S, 5)) return;  //Specify the number of input
ports that a block has.
   ssSetInputPortWidth(S, 0, 1);      //Specify the width of an input port.
   ssSetInputPortWidth(S, 1, 1);      //Specify the width of an input port.
   ssSetInputPortWidth(S, 2, 1);      //Specify the width of an input port.
   ssSetInputPortWidth(S, 3, 1);      //Specify the width of an input port.
   ssSetInputPortWidth(S, 4, 1);      //Specify the width of an input port.
   ssSetInputPortRequiredContiguous(S, 0, true); /*direct input signal
access*/
   ssSetInputPortRequiredContiguous(S, 1, true); /*direct input signal
access*/
   ssSetInputPortRequiredContiguous(S, 2, true); /*direct input signal
access*/
   ssSetInputPortRequiredContiguous(S, 3, true); /*direct input signal
access*/
```

```
    ssSetInputPortRequiredContiguous(S, 4, true); /*direct input signal
access*/
    /*
     * Set direct feedthrough flag (1=yes, 0=no).
     * A port has direct feedthrough if the input is used in either
     * the mdlOutputs or mdlGetTimeOfNextVarHit functions.
     * See matlabroot/simulink/src/sfuntmpl_directfeed.txt.
     */
    ssSetInputPortDirectFeedThrough(S, 0, 1); //Specify the direct feed-
through status of a block's ports.
    ssSetInputPortDirectFeedThrough(S, 1, 1); //Specify the direct feed-
through status of a block's ports.
    ssSetInputPortDirectFeedThrough(S, 2, 1); //Specify the direct feed-
through status of a block's ports.
    ssSetInputPortDirectFeedThrough(S, 3, 1); //Specify the direct feed-
through status of a block's ports.
    ssSetInputPortDirectFeedThrough(S, 4, 1); //Specify the direct feed-
through status of a block's ports.

    if (!ssSetNumOutputPorts(S, 2)) return;
    ssSetOutputPortWidth(S, 0, 1);
    ssSetOutputPortWidth(S, 1, 10);

    ssSetNumSampleTimes(S, 1);       //Specify the number of sample times that
an S-Function block has.
    ssSetNumRWork(S, 0);
    ssSetNumIWork(S, 0);
    ssSetNumPWork(S, 0);
    ssSetNumModes(S, 0);
    ssSetNumNonsampledZCs(S, 0);

    ssSetOptions(S, 0);
  }

  /* Function: mdlInitializeSampleTimes =================================
   * Abstract:
   *  This function is used to specify the sample time(s) for your
   *  S-function. You must register the same number of sample times as
   *  specified in ssSetNumSampleTimes.
   */
   static void mdlInitializeSampleTimes(SimStruct *S)
   {
```

```
  const real_T  *stime=mxGetPr(sampletime(S));
  ssSetSampleTime(S, 0, stime[0]);
  ssSetOffsetTime(S, 0, 0.0);
}

#define MDL_INITIALIZE_CONDITIONS /* Change to #undef to remove function
*/
#if defined(MDL_INITIALIZE_CONDITIONS)
/* Function: mdlInitializeConditions ================================
 * Abstract:
 *  In this function, you should initialize the continuous and discrete
 *  states for your S-function block. The initial states are placed
 *  in the state vector, ssGetContStates(S) or ssGetRealDiscStates(S).
 *  You can also perform any other initialization activities that your
 *  S-function may require. Note, this routine will be called at the
 *  start of simulation and if it is present in an enabled subsystem
 *  configured to reset states, it will be call when the enabled subsystem
 *  restarts execution to reset the states.
 */
static void mdlInitializeConditions(SimStruct *S)
{
}
#endif /* MDL_INITIALIZE_CONDITIONS */

#define MDL_START /* Change to #undef to remove function */
#if defined(MDL_START)
/* Function: mdlStart ==============================================
 * Abstract:
 *  This function is called once at start of model execution. If you
 *  have states that should be initialized once, this is the place
 *  to do it.
 */
static void mdlStart(SimStruct *S)
{
  count=0;
  label=0;
}
#endif /* MDL_START */

/* Function: mdlOutputs ============================================
 * Abstract:
 *  In this function, you compute the outputs of your S-function
```

```
* block. Generally outputs are placed in the output vector, ssGetY(S).
*/
static void mdlOutputs(SimStruct *S, int_T tid)
{
  const real_T *p = ssGetInputPortRealSignal(S,0);
  const real_T *dp = ssGetInputPortRealSignal(S,1);
  const real_T *v = ssGetInputPortRealSignal(S,2);
  const real_T *dv = ssGetInputPortRealSignal(S,3);
  const real_T *lu = ssGetInputPortRealSignal(S,4);
  const real_T *stime = mxGetPr(sampletime(S));
  real_T     *u = (real_T *)ssGetOutputPortSignal(S,0);
  real_T     *wi = (real_T *)ssGetOutputPortSignal(S,1);

  real_T Set_A[7] = {-1,-1,0,0,0,1,1};
  real_T Set_B[7] = {-1,-1,0,0,0,1,1};
  real_T Set_C[7] = {-1,-1,0,0,0,1,1};
  real_T Set_D[7] = {-1,-1,0,0,0,1,1};
  real_T Set_E[7] = {0,0,0.5,0.5,0.5,1,1};
  real_T Set_F[7] = {0,0,0.5,0.5,0.5,1,1};
  real_T A_p[3]  = {0,0,0};
  real_T A_dp[3] = {0,0,0};
  real_T A_v[3]  = {0,0,0};
  real_T A_dv[3] = {0,0,0};
  real_T A_ap[3] = {0,0,0};
  real_T A_av[3] = {0,0,0};
  real_T du_p[3] = {-1,0,1};
  real_T du_dp[3] = {-1,0,1};
  real_T du_v[3] = {-1,0,1};
  real_T du_dv[3] = {-1,0,1};
  real_T dw_p[3] = {0,0.5,1};
  real_T dw_dp[3] = {0,0.5,1};
  real_T dw_v[3] = {0,0.5,1};
  real_T dw_dv[3] = {0,0.5,1};
  real_T wB[4]  = {0,0,0,0};
  real_T B[4]  = {1,1,1,1};
  real_T du0[4],dw0[4],du,wD[4];
  real_T tmp_p,tmp_dp,tmp_v,tmp_dv,tmp_ap,tmp_av,sumdu;
real_T sumdu1[4],sumdu2[4],sumdw1[4],sumdw2[4],tmp_A1[3];
real_T tmp_du[3],tmp_A2[3],tmp_dw[3];
  int_T i,j,m;

  static real_T error1,error2;
  real_T xi[4],xo[8];
```

```
static int_T jj;
xi[0]=p[0];
xi[1]=dp[0];
xi[2]=v[0];
xi[3]=dv[0];
if (count==0)
{
  initialize_model_modification();
  error1=accumulate_control_error(xi,count);
  reset_gaussian_vector_center();
}
if(label==0)
{
switch(nostep)
{
  case 0:
    error1=accumulate_control_error(xi,count);
    keep_current_parameters();
    generate_gaussian_vector();
    modify_model_positively();
    for (i=0;i<4;i++)
    {
      xo[i]=tabweight[i];
      xo[i+4]=wchangesf[i];
    }
    nostep=1;
    break;
  case 1:
    error2=accumulate_control_error(xi,count);
    if(error2<0.0001)
    {
      label=1;
          for (i=0;i<4;i++)
          {
            xo[i]=tabweight[i];
            xo[i+4]=wchangesf[i];
          }
        for(i=0;i<3;i++)
          printf("error1=%f,%d\n",error1,count);
        for(i=0;i<4;i++)
          printf("xo[%d]=%f\n",i,tabweight[i]);
    break;
```

```
    }
    if (error2<error1)
    {
      jj=0;
      error1=error2;
      lengthen_vector_center();
      for (i=0;i<4;i++)
      {
        xo[i]=tabweight[i];
        xo[i+4]=wchangesf[i];
      }
      nostep=0;
      break;
    }
    else
    {
      restore_previous_model();
      modify_model_negatively();

      for (i=0;i<4;i++)
      {
        xo[i]=tabweight[i];
        xo[i+4]=wchangesf[i];
      }

      nostep=2;
    break;
    }
  case 2:
    error2=accumulate_control_error(xi,count);
    if(error2<0.0001)
    {
      label=1;
          for (i=0;i<4;i++)
          {
            xo[i]=tabweight[i];
            xo[i+4]=wchangesf[i];
          }
        for(i=0;i<3;i++)
          printf("error1=%f,%d\n",error1,count);
        for(i=0;i<4;i++)
          printf("xo[%d]=%f\n",i,tabweight[i]);
```

```
    break;
    }
    if (error2<error1)
    {
      jj=0;
      error1=error2;
      reverse_vector_center();
      for (i=0;i<4;i++)
      {
        xo[i]=tabweight[i];
        xo[i+4]=wchangesf[i];
      }
      nostep=0;
      break;
    }
    else
    {
      jj+=1;
      restore_previous_model();
      shorten_vector_center();

      for (i=0;i<4;i++)
      {
        xo[i]=tabweight[i];
        xo[i+4]=wchangesf[i];
      }

      nostep=2;

      if(jj>10)
      {
        jj=0;
        reset_gaussian_vector_center();
        for (i=0;i<4;i++)
        {
          xo[i]=tabweight[i];
          xo[i+4]=wchangesf[i];
        }
        nostep=0;
      }
    }
  }
}
```

```
  else
{
        for (i=0;i<4;i++)
      {
        xo[i]=tabweight[i];
        xo[i+4]=wchangesf[i];
      }
}
wi[8]=error1;
wi[9]=error2;
count=count+1;

////// The End of The Forth Table //////
  tmp_p = p[0];
  tmp_dp = dp[0];
  tmp_v = v[0];
  tmp_dv = dv[0];
  tmp_ap = fabs(p[0]);
  tmp_av = fabs(v[0]);
  for (i=0;i<4;i++)
  {
    wB[i]=xo[i];
    B[i]=xo[4+i];
  }

////// The First Table: p //////
  if (tmp_p<Set_A[0])
  A_p[0]=1;
  else if (tmp_p>=Set_A[0]&&tmp_p<=Set_A[2])
  A_p[0]=-1/(Set_A[2]-Set_A[0])*(tmp_p-Set_A[0])+1;
  else;
  if (tmp_p>=Set_A[1]&&tmp_p<Set_A[3])
  A_p[1]=1/(Set_A[3]-Set_A[1])*(tmp_p-Set_A[3])+1;
  else if (tmp_p>=Set_A[3]&&tmp_p<Set_A[5])
  A_p[1]=-1/(Set_A[5]-Set_A[3])*(tmp_p-Set_A[3])+1;
  else;
  if (tmp_p>=Set_A[4]&&tmp_p<Set_A[6])
  A_p[2]=1/(Set_A[6]-Set_A[4])*(tmp_p-Set_A[6])+1;
  else if (tmp_p>=Set_A[6])
  A_p[2]=1;
  else;
////// The End of The First Table //////
```

```
////// The Second Table: dp //////
 if (tmp_dp<Set_B[0])
 A_dp[0]=1;
 else if (tmp_dp>=Set_B[0]&&tmp_dp<=Set_B[2])
 A_dp[0]=-1/(Set_B[2]-Set_B[0])*(tmp_dp-Set_B[0])+1;
 else;
 if (tmp_dp>=Set_B[1]&&tmp_dp<Set_B[3])
 A_dp[1]=1/(Set_B[3]-Set_B[1])*(tmp_dp-Set_B[3])+1;
 else if (tmp_dp>=Set_B[3]&&tmp_dp<Set_B[5])
 A_dp[1]=-1/(Set_B[5]-Set_B[3])*(tmp_dp-Set_B[3])+1;
 else;
 if (tmp_dp>=Set_B[4]&&tmp_dp<Set_B[6])
 A_dp[2]=1/(Set_B[6]-Set_B[4])*(tmp_dp-Set_B[6])+1;
 else if (tmp_dp>=Set_B[6])
 A_dp[2]=1;
 else;
////// The End of The Second Table //////

////// The Third Table: v //////
 if (tmp_v<Set_C[0])
 A_v[0]=1;
 else if (tmp_v>=Set_C[0]&&tmp_v<=Set_C[2])
 A_v[0]=-1/(Set_C[2]-Set_C[0])*(tmp_v-Set_C[0])+1;
 else;
 if (tmp_v>=Set_C[1]&&tmp_v<Set_C[3])
 A_v[1]=1/(Set_C[3]-Set_C[1])*(tmp_v-Set_C[3])+1;
 else if (tmp_v>=Set_C[3]&&tmp_v<Set_C[5])
 A_v[1]=-1/(Set_C[5]-Set_C[3])*(tmp_v-Set_C[3])+1;
 else;
 if (tmp_v>=Set_C[4]&&tmp_v<Set_C[6])
 A_v[2]=1/(Set_C[6]-Set_C[4])*(tmp_v-Set_C[6])+1;
 else if (tmp_v>=Set_C[6])
 A_v[2]=1;
 else;
////// The End of The Third Table //////

////// The Fouth Table: dv //////
 if (tmp_dv<Set_D[0])
 A_dv[0]=1;
 else if (tmp_dv>=Set_D[0]&&tmp_dv<=Set_D[2])
 A_dv[0]=-1/(Set_D[2]-Set_D[0])*(tmp_dv-Set_D[0])+1;
 else;
```

```
  if (tmp_dv>=Set_D[1]&&tmp_dv<Set_D[3])
  A_dv[1]=1/(Set_D[3]-Set_D[1])*(tmp_dv-Set_D[3])+1;
  else if (tmp_dv>=Set_D[3]&&tmp_dv<Set_D[5])
  A_dv[1]=-1/(Set_D[5]-Set_D[3])*(tmp_dv-Set_D[3])+1;
  else;
  if (tmp_dv>=Set_D[4]&&tmp_dv<Set_D[6])
  A_dv[2]=1/(Set_D[6]-Set_D[4])*(tmp_dv-Set_D[6])+1;
  else if (tmp_dv>=Set_D[6])
  A_dv[2]=1;
  else;
////// The End of The Fouth Table //////

////// The Fifth Table: ap //////
  if (tmp_ap>=Set_E[0]&&tmp_ap<=Set_E[2])
  A_ap[0]=-1/(Set_E[2]-Set_E[0])*(tmp_ap-Set_E[0])+1;
  else;
  if (tmp_ap>=Set_E[1]&&tmp_ap<Set_E[3])
  A_ap[1]=1/(Set_E[3]-Set_E[1])*(tmp_ap-Set_E[3])+1;
  else if (tmp_ap>=Set_E[3]&&tmp_ap<Set_E[5])
  A_ap[1]=-1/(Set_E[5]-Set_E[3])*(tmp_ap-Set_E[3])+1;
  else;
  if (tmp_ap>=Set_E[4]&&tmp_ap<Set_E[6])
  A_ap[2]=1/(Set_E[6]-Set_E[4])*(tmp_ap-Set_E[6])+1;
  else if (tmp_ap>=Set_E[6])
  A_ap[2]=1;
  else;
////// The End of The Fifth Table //////

////// The Sixth Table: av //////
  if (tmp_av>=Set_F[0]&&tmp_av<=Set_F[2])
  A_av[0]=-1/(Set_F[2]-Set_F[0])*(tmp_av-Set_F[0])+1;
  else;
  if (tmp_av>=Set_F[1]&&tmp_av<Set_F[3])
  A_av[1]=1/(Set_F[3]-Set_F[1])*(tmp_av-Set_F[3])+1;
  else if (tmp_av>=Set_F[3]&&tmp_av<Set_F[5])
  A_av[1]=-1/(Set_F[5]-Set_F[3])*(tmp_av-Set_F[3])+1;
  else;
  if (tmp_av>=Set_F[4]&&tmp_av<Set_F[6])
  A_av[2]=1/(Set_F[6]-Set_F[4])*(tmp_av-Set_F[6])+1;
  else if (tmp_av>=Set_F[6])
  A_av[2]=1;
  else;
////// The End of The Sixth Table //////
```

```
////// Simplified Fuzzy Inference ///////
  sumdu=0;
  for (j=0;j<4;j++)
  {
    sumdu1[j]=0;
    sumdu2[j]=0;
    sumdw1[j]=0;
    sumdw2[j]=0;
    switch (j)
    {
      case 0:
        for (m=0;m<3;m++)
        {
          tmp_A1[m]=A_p[m];
          tmp_du[m]=du_p[m];
          tmp_A2[m]=A_ap[m];
          tmp_dw[m]=dw_p[m];
        }
        break;
      case 1:
        for (m=0;m<3;m++)
        {
          tmp_A1[m]=A_dp[m];
          tmp_du[m]=du_dp[m];
          tmp_A2[m]=A_ap[m];
          tmp_dw[m]=dw_dp[m];
        }
        break;
      case 2:
        for (m=0;m<3;m++)
        {
          tmp_A1[m]=A_v[m];
          tmp_du[m]=du_v[m];
          tmp_A2[m]=A_av[m];
          tmp_dw[m]=dw_v[m];
        }
        break;
      case 3:
        for (m=0;m<3;m++)
        {
          tmp_A1[m]=A_dv[m];
          tmp_du[m]=du_dv[m];
```

```
            tmp_A2[m]=A_av[m];
            tmp_dw[m]=dw_dv[m];
          }
          break;
      }

      for (i=0;i<3;i++)
      {
        sumdu1[j]=sumdu1[j]+tmp_A1[i]*tmp_du[i];
        sumdu2[j]=sumdu2[j]+tmp_A1[i];

        sumdw1[j]=sumdw1[j]+tmp_A2[i]*tmp_dw[i];
        sumdw2[j]=sumdw2[j]+tmp_A2[i];
      }
      du0[j]=sumdu1[j]/sumdu2[j];
      dw0[j]=sumdw1[j]/sumdw2[j];
      wD[j]=wB[j]+B[j]*dw0[j];
      sumdu=sumdu+wD[j]*du0[j];
      wi[j]=wB[j];
      wi[4+j]=B[j];
    }

    du=sumdu;
    u[0]=lu[0]/1+du;
  }

  #define MDL_UPDATE /* Change to #undef to remove function */
  #if defined(MDL_UPDATE)
 /* Function: mdlUpdate ==============================================
  * Abstract:
  *  This function is called once for every major integration time step.
  *  Discrete states are typically updated here, but this function is useful
  *  for performing any tasks that should only take place once per
  *  integration step.
  */
  static void mdlUpdate(SimStruct *S, int_T tid)
  {
  }
  #endif /* MDL_UPDATE */

  #define MDL_DERIVATIVES /* Change to #undef to remove function */
  #if defined(MDL_DERIVATIVES)
```

```
/* Function: mdlDerivatives =========================================
 * Abstract:
 *  In this function, you compute the S-function block's derivatives.
 *  The derivatives are placed in the derivative vector, ssGetdX(S).
 */
 static void mdlDerivatives(SimStruct *S)
 {
 }
 #endif /* MDL_DERIVATIVES */

/* Function: mdlTerminate ===========================================
 * Abstract:
 *  In this function, you should perform any actions that are necessary
 *  at the termination of a simulation. For example, if memory was
 *  allocated in mdlStart, this is the place to free it.
 */
 static void mdlTerminate(SimStruct *S)
 {
 }

/*=============================================================*
 * See sfuntmpl_doc.c for the optional S-function methods *
 *=============================================================*/

/*===============================*
 * Required S-function trailer *
 *===============================*/

 #ifdef MATLAB_MEX_FILE  /* Is this file being compiled as a MEX-file? */
 #include "simulink.c"   /* MEX-file interface mechanism */
 #else
 #include "cg_sfun.h"   /* Code generation registration function */
 #endif

 ///////////////////////////File funRSde.h//////////////////////////
 #include "math.h"
 #include "stdlib.h"
 #include "minmax.h"
 static  double  tunepara;
 static  double  tabweight[4];       /* base value of dynamic importance
degrees */
 static  double  wchangesf[4];       /* changing breadth of dynamic impor-
tance degrees */
```

```
   static  double  inc_baseval[4];        /* increment amount for base values */
   static  double  old_baseval[4];        /* previous value for base values */
   static  double  gv_baseval[4];         /* Gaussian vector for base values */
   static  double  inc_breadth[4];   /*  increment  amount  for  changing
breadths */
   static  double  old_breadth[4];    /*  previous  value  for  changing
breadths */
   static  double  gv_breadth[4];     /*  Gaussian  vector  for  changing
breadths */
   static  double  past_error_var[20];

   void initialize_model_modification(void);
   void reset_gaussian_vector_center(void);
   void generate_gaussian_vector(void);
   void keep_current_parameters(void);
   void modify_model_positively(void);
   void modify_model_negatively(void);
   void restore_previous_model(void);
   void lengthen_vector_center(void);
   void reverse_vector_center(void);
   void shorten_vector_center(void);
   double accumulate_control_error(double *,int);
   void start_random_search_procedure(void);

   /*--------------------------------------------------------------
     Function Name: initialize_model_modification(void)
     Return Value : void
     Main Contents: initialize the modification amount
   ---------------------------------------------------------------*/
   void initialize_model_modification(void)
   {
     int     k;                  /* count */
     for (k = 0; k < 4; k++) {
       inc_baseval[k] = 0.0;
       inc_breadth[k] = 0.0;
     }
   }

   /*--------------------------------------------------------------
     Function Name: reset_gaussian_vector_center(void)
     Return Value : void
     Main Contents: randomly reset Gaussian vector centers for random serach
   --------------------------------------------------------------*/
```

```
void reset_gaussian_vector_center(void)
{
  int     k;                  /* count */
  for (k = 0; k < 4; k++) {
    gv_baseval[k] = ((double)rand() / 65535.0) * 100.0;
    gv_breadth[k] = ((double)rand() / 65535.0) * 100.0;
  }
}

/*------------------------------------------------------------------
  Function Name: generate_gaussian_vector(void)
  Return Value : void
  Main Contents: generate new Gaussian vectors with the centers, each
          component of which is independent and has the same variance
-------------------------------------------------------------------*/
void generate_gaussian_vector(void)
{
  int     k;                  /* count */
  double  idx, idy;
  double  norm1 = 1.0 / 32768.01;
  double  norm2 = 2.0 * 3.1415926 / 32768.01;
  tunepara=1;
  for (k = 0; k < 4; k++) {
    idx = (double)rand() * norm1 + norm1;
    idy = (double)rand() * norm2 + norm2;
    inc_baseval[k] = gv_baseval[k] + sqrt(-2.0 * log10(idx)) * cos(idy) *
tunepara;
  }
  for (k = 0; k < 4; k++) {
    idx = (double)rand() * norm1 + norm1;
    idy = (double)rand() * norm2 + norm2;
    inc_breadth[k] = gv_breadth[k] + sqrt(-2.0 * log10(idx)) * cos(idy) *
tunepara;
  }
}

/*------------------------------------------------------------------
  Function Name: keep_current_parameters(void)
  Return Value : void
  Main Contents: temporarily keep the current set of the DID parameters
-------------------------------------------------------------------*/
void keep_current_parameters(void)
```

```
{
  int     k;                  /* count */
  for (k = 0; k < 4; k++) {
    old_baseval[k] = tabweight[k];
    old_breadth[k] = wchangesf[k];
  }
}

/*------------------------------------------------------------------
  Function Name: modify_model_positively(void)
  Return Value : void
  Main Contents: modify the DID parameters in positive direction
          after each random searching iteration
------------------------------------------------------------------*/
void modify_model_positively(void)
{
  int     k;                  /* count */
  for (k = 0; k < 4; k++) {
    tabweight[k] += inc_baseval[k];
    tabweight[k] = max(0.0, tabweight[k]);
    wchangesf[k] += inc_breadth[k];
    wchangesf[k] = max(0.0, wchangesf[k]);
  }
}

/*------------------------------------------------------------------
  Function Name: modify_model_negatively(void)
  Return Value : void
  Main Contents: modify the DID parameters in negative direction
          after each random searching iteration
------------------------------------------------------------------*/
void modify_model_negatively(void)
{
  int     k;                  /* count */
  for (k = 0; k < 4; k++) {
    tabweight[k] -= inc_baseval[k];
    tabweight[k] = max(0.0, tabweight[k]);
    wchangesf[k] -= inc_breadth[k];
    wchangesf[k] = max(0.0, wchangesf[k]);
  }
}
```

```
/*------------------------------------------------------------------
  Function Name: restore_previous_model(void)
  Return Value : void
  Main Contents: restore all the parameters to the previous set
------------------------------------------------------------------*/
void restore_previous_model(void)
{
  int     k;                  /* count */
  for (k = 0; k < 4; k++) {
    tabweight[k] = old_baseval[k];
    wchangesf[k] = old_breadth[k];
  }
}

/*------------------------------------------------------------------
  Function Name: lengthen_vector_center(void)
  Variables    : void
  Main Contents: lengthen the Gaussian vector centers in positive direction
------------------------------------------------------------------*/
void lengthen_vector_center(void)
{
  int     k;                  /* count */
  for (k = 0; k < 4; k++) {
    gv_baseval[k] = 0.2 * gv_baseval[k] + 0.4 * inc_baseval[k];
    gv_breadth[k] = 0.2 * gv_breadth[k] + 0.4 * inc_breadth[k];
  }
}

/*------------------------------------------------------------------
  Function Name: reverse_vector_center(void)
  Return Value : void
  Main Contents: reverse the Gaussian vector centers to negative direction
------------------------------------------------------------------*/
void reverse_vector_center(void)
{
  int     k;                  /* count */
  for (k = 0; k < 4; k++) {
    gv_baseval[k] -= 0.4 * inc_baseval[k];
    gv_breadth[k] -= 0.4 * inc_breadth[k];
  }
}
```

```
/*-------------------------------------------------------------------
  Function Name: shorten_vector_center(void)
  Return Value : void
  Main Contents: shorten the Gaussian vector centers to half their originals
-------------------------------------------------------------------*/
void shorten_vector_center(void)
{
  int     k;                /* count */
  for (k = 0; k < 4; k++) {
    gv_baseval[k] = 0.5 * gv_baseval[k];
    gv_breadth[k] = 0.5 * gv_breadth[k];
  }
}

/*-------------------------------------------------------------------
  Function Name: accumulate_control_error(void)
  Return Value : double      return total absolute error
  Main Contents: execute fuzzy control based on the SIRMs dynamically
          connected fuzzy inference model
          at the same time, accumulate the absolute error
          finally return the total absolute error as its performance
-------------------------------------------------------------------*/
double accumulate_control_error(double * errorvar,int count)
{
  double   error1 = 0.0;          /* total absolute error */
  double sum,error;
  int i;
  error1 += (fabs(errorvar[0])
     + fabs(errorvar[1])
     + fabs(errorvar[2])
     + fabs(errorvar[3]))
     ;
  for(i=19;i>0;i--)
    past_error_var[i]=past_error_var[i-1];
  past_error_var[0]=error1;
  if (count>=19)
  {
    sum=0;
    for(i=0;i<20;i++)
      sum=sum+past_error_var[i];
    error=sum/20;
  }
```

```
    else
    {
      /* accumulate total absolute error */
      sum=0;
      for(i=0;i<20;i++)
      sum=sum+past_error_var[i];
      error=sum/(count+1);
    }

    return (error);
}
```

4) Program running result (DID, Version 2)

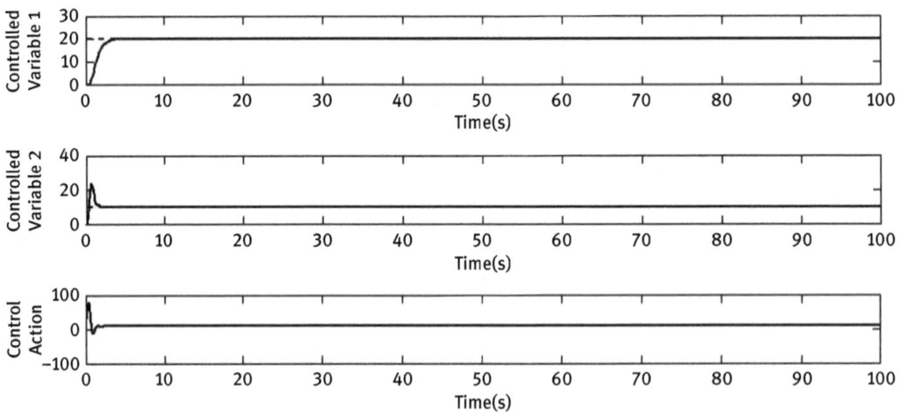

Figure 5.29: Simulative results of the SIRMs-based fuzzy control with DID (Version 2).

5.5.4 Output tracking control

1) Fuzzy Cluster Modeling for the Hydrogen-Air PEM Fuel Cell System

Note: The programs for the modeling of the hydrogen-air PEM fuel cell system are based on the work of Robert Babuška. Readers can download the FMID 4.0 Toolbox from his website https://d1rkab7tlqy5f1.cloudfront.net/3mE/Organisatie/Afdelingen/Delft%20Center%20for%20Systems%20and%20Control/People/Personal%20Pages/Robert%20Babuska/fmid-v40.zip. The only difference is that we modified the file 'fmsim.m' to obtain a special form of model. For the readers' convenience, we pasted the relevant source codes.

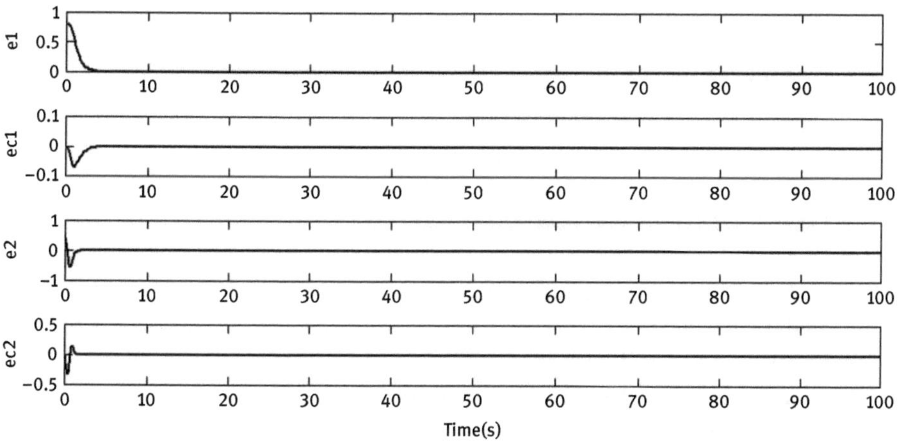

Figure 5.30: Errors and change of errors in the SIRMs-based fuzzy control with DID (Version 2).

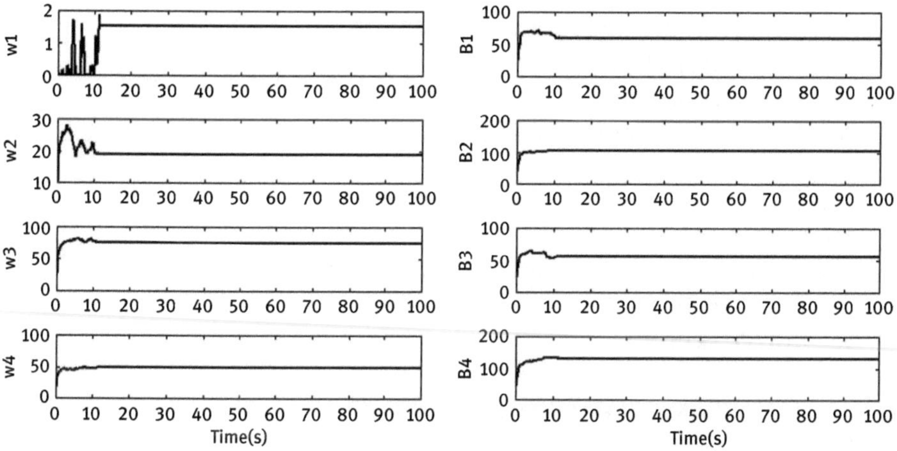

Figure 5.31: Optimized parameters: base value w_i and breadth B_i, where i=1,2,3,4 (Version 2).

```
% generate a fuzzy model for the hydrogen-air pem fuel cell system
clear FM
%%%%%%%%%%%%%%%%%%%%%%%%%%%%%%%%%%%%%%%%%
% define constants
%%%%%%%%%%%%%%%%%%%%%%%%%%%%%%%%%%%%%%%%%
FM.c =3;          % number of clusters5
FM.m = 5;         % fuzziness parameter4.5
FM.tol = 0.01;    % termination criterion
FM.seed =0;       % seed
FM.ante = 1;      % antecedent: 1 - product-space MFS
```

```
                    %                 2 - projected MFS
FM.Ny = 1;          % denominator order
FM.Nu = [1];        % numerator orders
FM.Nd = [1];        % transport delays
                    % (set to 1 for y(k+1) = f(u(k),....))

%%%%%%%%%%%%%%%%%%%%%%%%%%%%%%%%%%%%%%%%
% read data
%%%%%%%%%%%%%%%%%%%%%%%%%%%%%%%%%%%%%%%%
load data20150607
skip = 1;             % decimation
N = size(y,1);          % split data in halves
N2 = floor(size(y,1)/2);

%%%%%%%%%%%%%%%%%%%%%%%%%%%%%%%%%%%%%%%%
% identification data
%%%%%%%%%%%%%%%%%%%%%%%%%%%%%%%%%%%%%%%%
Dat.U = u(1:skip:N2);
Dat.Y = y(1:skip:N2);
Dat.Ts =0.1;     % sample time [s]

%%%%%%%%%%%%%%%%%%%%%%%%%%%%%%%%%%%%%%%%
% validation data
%%%%%%%%%%%%%%%%%%%%%%%%%%%%%%%%%%%%%%%%
ue = u(N2+1:skip:N);
ye = y(N2+1:skip:N);

%%%%%%%%%%%%%%%%%%%%%%%%%%%%%%%%%%%%%%%%
% make fuzzy model by means of fuzzy clustering
%%%%%%%%%%%%%%%%%%%%%%%%%%%%%%%%%%%%%%%%
[FM,Part] = fmclust1(Dat,FM);

%%%%%%%%%%%%%%%%%%%%%%%%%%%%%%%%%%%%%%%%
% simulate the fuzzy model for validation data
%%%%%%%%%%%%%%%%%%%%%%%%%%%%%%%%%%%%%%%%
figure(1); clf
[ym,VAF,dof,yl,ylm] = fmsim1(ue,ye,FM,[],[],[],1); VAF
title('Process output (blue) and model output (magenta)');
ylabel('Output');
hold on
subplot(312); plot([ylm{1}]);
title('Individual local models');
```

```
xlabel('Time'); ylabel('Output');
subplot(313); plot(dof{1})
title('Degrees of fulfillment');
xlabel('Time'); ylabel('Membership grade');
save FM20150607 FM
fm2tex(FM,'fm20150607.tex')
disp('Hit any key to see the membership functions ...'); pause

plotmfs(FM);
disp('Hit any key to finish ...'); pause

close(1);

clc

%%%%%%%%%%%%%%%%%%%%%%%fmsim1.m%%%%%%%%%%%
function [Ym,q,DOF,Yl,Ylm] = fmsim1(Ue,Ye,FM,Ymin,Ymax,show,H)
% FMSIM simulate a MIMO fuzzy model.
%
% [Ym,q,DOF,Yl,Ylm] = FMSIM(U,Y,FM,Ymin,Ymax,show,H)
%
%   U,Y ... input and output data matrices, respectively
%        (output data is needed for initialization
%        and for comparison)
%   FM ... structure containing fuzzy model parameters
%   Ymin ... lower bound constraint on Y (optional)
%   Ymax ... upper bound constraint on Y (optional)
%   show ... set to 1 for on-line graphics, 2 for final plot only,
%    and 0 for no graphics (optional, default 1)
%   H  ... prediction horizon (optional, by default simulation)
%
%   Ym ... simulated output of the model
%   q ... performance index of the model computed as VAF
%       (variance accounted for)
%   DOF ... degrees of fulfillment of the rules (all outputs
%       appended)
%   Yl .... output of the individual rule's consequents
%   Ylm ... output of the individual rule's consequents
%       (data with DOF < 0.5 are masked by NaN)
%
% See also FMCLUST, FMSTRUCT, GKFAST, VAF.

% (c) Robert Babuska, Stanimir Mollov 1997-1999, Modified by TongSW(2015)
%
```

```
%# scalar i j kk k nset

if ~iscell(Ue), Ue = {Ue}; end; if ~iscell(Ye), Ye = {Ye}; end;

%Update an old-type model
FM = fmupdate(FM);

for nset = 1 : length(Ue),

U = Ue{nset};
Y = Ye{nset};

%%%%%%%%%%%%%%%%%%%%%%%%%%%%%%%%%%%%%%%
% extract data from the FM structure (much faster!)
%%%%%%%%%%%%%%%%%%%%%%%%%%%%%%%%%%%%%%%
Ninps = size(U,2);
NO = FM.no;

ny = FM.Ny;
nu = FM.Nu;

atype = FM.ante;
rls = FM.rls;
mfs = FM.mfs;
p = FM.th;

for k = 1 : NO,
  c(k) = size(p{k},1);
  V1{k} = ones(c(k),1);
end;

if any(atype == 1),
  V = FM.V;
  M = FM.M;
  m = FM.m;
end;

%NI = sum(ny,2) + sum(nu,2); % Robert
NI = max(size([ny{1}{:}],2), size([ny{2}{:}],2)) + max(size([nu{1}
{:}],2), size([nu{2}{:}],2)); %Stanimir
if isempty(Y), Y = zeros(size(U,1),NO); end;
if nargin < 4, Ymin = -inf*ones(1,NO); elseif isempty(Ymin), Ymin =
-inf*ones(1,NO); end;
```

```
    if nargin < 5, Ymax = inf*ones(1,NO); elseif isempty(Ymax), Ymax = inf*ones
 (1,NO); end;
    if nargin < 6, show = 1; elseif isempty(show), show = 1; end;
    if nargin < 7, H = 0; elseif isempty(H), H = 0; end;

    %%%%%%%%%%%%%%%%%%%%%%%%%%%%%%%%%%%%%%%%
    % intialize graphics
    %%%%%%%%%%%%%%%%%%%%%%%%%%%%%%%%%%%%%%%%
    if show == 1,
    clf;
    if ~((max(size([ny{1}{:}],2))|max(size([ny{2}{:}],2))) == 0| NI == 1 | H~=0),
      Ts = FM.Ts;
     for i = 1:NO,
      subplot(NO,1,i);
      plot(Ts*(1:length(Y(:,i))),Y(:,i));
      set(gca,'xlim',Ts*[1 size(Y,1)]);
      xlabel('Time [s]'); ylabel(FM.OutputName{i}); drawnow; hold on;
     end;
   end;
  end;

   %%%%%%%%%%%%%%%%%%%%%%%%%%%%%%%%%%%%%%%%
   % static model
   %%%%%%%%%%%%%%%%%%%%%%%%%%%%%%%%%%%%%%%%
   %if max(max(ny)) == 0, %Robert
   if (max(size([ny{1}{:}],2))+max(size([ny{2}{:}],2))) == 0,  %Stanimir
   for kk = 1 : NO,

     %antecedent regressor
     [trash,ante,trash]=regres([ones(size(U,1),1) U],ny{1}(k,:),nu{1}
(k,:));
     x1 = ones(size(ante,1),1);
     %consequent regressor
     [trash,consq,trash]=regres([ones(size(U,1),1) U],ny{2}(k,:),nu{2}
(k,:));
     x2 = ones(size(consq,1),1);
     if isempty(x2), x2 = x1;, end
     cons = [consq x2];

     if ~isempty(FM.th{kk}),
       if c(kk) > 1,
        if atype(kk) == 1,  % product-space MFs
         for j = 1 : c(kk),            % for all clusters
```

```
          xv = ante - x1*V{kk}(j,:);
          d(:,j) = sum(xv*M{kk}(:,:,j).*xv,2);
        end;
        d = (d+1e-100).^(-1/(m(kk)-1));   % clustering dist
        DOF{kk} = (d ./ (sum(d,2)*ones(1,c(kk))));
       elseif atype(kk) == 2, % projected MFs
        DOF{kk} = dofprod(FM.rls{kk},FM.mfs{kk},ante);
       end;
       [Ym(:,kk),Yl{kk},Ylm{kk}] = sugval(FM.th{kk},cons,DOF{kk});
      else
       Ym(:,kk) = cons*FM.th{kk}';
       DOF{kk} = ones(size(Ym(:,kk)));
       Yl{kk} = Ym(:,kk);
       Ylm{kk} = Yl{kk};
      end;
    else
      Ym(:,kk) = Y(end-size(ante,1)+1:end,kk);
      DOF{kk} = NaN*Ym(:,kk);
      Yl{kk} = DOF{kk};
      Ylm{kk} = Yl{kk};
    end;
    Y = Y(end-size(ante,1)+1:end,:);
  end;
  elseif H > 0, % dynamic model, H-step-ahead simulation

  N = size(Y,1);
  Ym = Y;

  for kk = 1 : NO,          % for all outputs
  if Ninps == 0, z = Y; else z = [Y U]; end;

  %antecedent regressor%
  [yr,ante,y1]=regres([Y(:,kk) z],ny{1}(kk,:),nu{1}(kk,:));
  x1 = ones(size(ante,1),1);

  %consequent regressor
  [yr,consq,y1]=regres([Y(:,kk) z],ny{2}(kk,:),nu{2}(kk,:));
  x2 = ones(size(consq,1),1);
  %  cons = [consq x2];
  cons=[consq];%Tong

  if ~isempty(FM.th{kk}),
    if c(kk) > 1,
```

```
        if atype(kk) == 1,   % product-space MFs
         d = [];
         for j = 1 : c(kk),            % for all clusters
          xv = ante - x1*V{kk}(j,:);
          d(:,j) = sum(xv*M{kk}(:,:,j).*xv,2);
         end;
         d = (d+1e-100).^(-1/(m(kk)-1));   % clustering dist
         dofe = (d ./ (sum(d,2)*ones(1,c(kk))));
         % dofe = 1./(1+d);        % similarity
        elseif atype(kk) == 2, % projected MFs
         dofe = dofprod(rls{kk},mfs{kk},ante);
        end;
        s1 = size(ante,1);
        [Ym(N-s1+1:N,kk),Yl{kk}(N-s1+1:N,:),Ylm{kk}(N-s1+1:N,:)] = sugval
(p{kk},cons,dofe);
        DOF{kk} = dofe;
      else
        s1 = size(ante,1);
        Ym(N-s1+1:N,kk) = cons*p{kk}';
        DOF{kk} = 1;
      end;
    else
      Ym(:,kk) = Y(:,kk);
      DOF{kk} = NaN*Ym(:,kk);
      Yl{kk} = DOF{kk};
      Ylm{kk} = Yl{kk};
      end;
    end;

    else  % dynamic model, simulation from input
    if Ninps > 0,

      k0 = fmorder(ny,nu);

      if size(U,1) < k0, error(['Supply at least ' int2str(k0) ' data samples.']);
end;
    else
      k0 = max(max(max([ny{1}{:}]),max([ny{2}{:}]))+1,2);
    end;
    if Ninps > 0, kmax = size(U,1); else kmax = size(Y,1); end;

    Ym = Y;
    for k = k0 : kmax,              % discrete time
```

```
for kk = 1 : NO,              % for all ouputs
 if ~isempty(FM.th{kk}),
  ante = [];
  cons = [];
  for j = 1 : NO,
   for s = 1:size(ny{1}{kk,j},2)
    pos = ny{1}{kk,j}(s);
    if isempty(pos), pos = 0; end
    ante = [ante Ym(k-pos,j)'];
   end
   for s = 1:size(ny{2}{kk,j},2)
    pos = ny{2}{kk,j}(s);
    if isempty(pos), pos = 0; end
    cons = [cons Ym(k-pos,j)'];
   end
  end;
  for j = 1 : Ninps,
   for s = 1:size(nu{1}{kk,j},2)
    ante = [ante U(k-nu{1}{kk,j}(s),j)'];
   end
   for s = 1:size(nu{2}{kk,j},2)
    cons = [cons U(k-nu{2}{kk,j}(s),j)'];
   end
  end;
  %%cons = [cons 1];
  cons=[cons]; % Tong Modified.
  if c(kk) > 1,
   if atype(kk) == 1,   % product-space MFs
    %    ante = max(min(ante,FM.zmax{kk}(1:NI)),FM.zmin{kk}(1:NI));
    xv = V1{kk}*ante - V{kk};
    d = [];
    for j = 1 : c(kk),
     d(j) = xv(j,:)*M{kk}(:,:,j)*xv(j,:)';
    end;
    d = (d+1e-100).^(-1/(m(kk)-1));   % clustering distance
    dofe = (d ./ (sum(d,2)*ones(1,c(kk))));
    %  dofe = 1./(1+d);          % similarity
   elseif atype(kk) == 2, % projected MFs
    dofe = dofprod(rls{kk},mfs{kk},ante);
   end;
   yl = cons*p{kk}';        % local models
   ylm = yl;
```

```
          ylm(find(dofe < max(dofe))) = NaN;   % mask with NaN's for plots
          ds = sum(dofe);
          if ds >0,
            Ym(k,kk) = yl*(dofe/ds)';
          else
            Ym(k,kk) = Ym(k-1,kk);
          end;
          Yl{kk}(k,:) = yl;
          Ylm{kk}(k,:) = ylm;
          DOF{kk}(k-1,:) = dofe;
        else
          Ym(k,kk) = cons*p{kk}';
          DOF{kk}(k-1,:) = 1;
          Yl{kk}(k,:) = Ym(k,kk);
          Ylm{kk}(k,:) = Ym(k,kk);
        end;
        if Ym(k,kk) < Ymin(kk),
          Ym(k,kk) = Ymin(kk);
        end;
        if Ym(k,kk) > Ymax(kk),
          Ym(k,kk) = Ymax(kk);
        end;
      else
        DOF{kk}(k-1,:) = NaN;
        Yl{kk}(k,:) = NaN;
        Ylm{kk}(k,:) = NaN;
      end;

      if show == 1,
        subplot(NO,1,kk);
         plot(Ts*[k-1 k],[Ym(k-1,kk) Ym(k,kk)],'m-','EraseMode','none');
drawnow;
      end;
    end;
  end;
end;

  %%%%%%%%%%%%%%%%%%%%%%%%%%%%%%%%%%%%%%%%
  % final plot
  %%%%%%%%%%%%%%%%%%%%%%%%%%%%%%%%%%%%%%%%
  %show = 0;
  if show,
```

```
    clf; hold off;
    for i = 1:NO,
      subplot(NO,1,i);
      if (max(size([ny{1}{:}],2))& max(size([ny{2}{:}],2))) == 0 &...
          (max(size([nu{1}{:}],2))&max(size([nu{2}{:}],2))) == 1 & NI == 1,
% static SISO model
subplot(3,1,1);
        plot(U,Y(:,i),'b-',U,Ym(:,i),'m-.');
        xlabel('Input'); ylabel(FM.OutputName{i}); drawnow;
      else % dynamic model
        Ts = FM.Ts;
    subplot(3,1,1);
          plot(Ts*(1:length(Y(:,i))),Y(:,i),'b-',Ts*(1:length(Ym(:,i))),Ym
(:,i),'m-.');
        xlabel('Time [s]'); ylabel(FM.OutputName{i}); drawnow;
      end;
     end;
    end;

    if size(Y,1)==size(Ym,1) & size(Ym,1)>1, q = vaf(Y,Ym); end;

    end;
```

2) Obtain the coefficients of the simplified model by neural network approximation
Note: We can use the neural network technology to approximate the membership functions of the antecedent variables for the fuzzy clustering model by the following steps.

Step1: Get the input-output data for each membership functions of the antecedent variables in the fuzzy clustering model.

Set break points in the file 'plotmfs.m', then excute the following codes.

```
if (k==1)
{
  yyy1=dom';
  aaa1=f(:,1)';
  bbb1=f(:,2)';
  ccc1=f(:,3)'
  save memfy1 yyy1 aaa1
  save memfy2 yyy1 bbb1
  save memfy3 yyy1 ccc1
}
else
```

```
{
  uuu2=dom';
  aaa2=f(:,1)';
  bbb2=f(:,2)';
  ccc2=f(:,3)'
  save memfu1 uuu2 aaa2
  save memfu2 uuu2 bbb2
  save memfu3 uuu2 ccc2
}
```

Step 2: Using the matlab toolbox 'nftool' to get the neural network approximation, named 'memfynet1', 'memfynet2', 'memfunet1','memfunet2' for the corresponding membership functions, we executed the following codes:

```
load FM20150607;

coefA=(sim(memfynet1.net,1)*sim(memfunet1.net,1)*FM.th{1}(1,1)+sim(mem-
fynet2.net,1)*sim(memfunet2.net,1)*FM.th{1}(2,1)+sim(memfynet3.net,1)
*sim(memfunet3.net,1)*FM.th{1}(3,1))/(sim(memfynet1.net,1)*sim(memfu-
net1.net,1)+sim(memfynet2.net,1)*sim(memfunet2.net,1)+sim(memfynet3.
net,1)*sim(memfunet3.net,1));

coefB=(sim(memfynet1.net,1)*sim(memfunet1.net,1)*FM.th{1}(1,2)+sim(mem-
fynet2.net,1)*sim(memfunet2.net,1)*FM.th{1}(2,2)+sim(memfynet3.net,1)
*sim(memfunet3.net,1)*FM.th{1}(3,2))/(sim(memfynet1.net,1)*sim(memfu-
net1.net,1)+sim(memfynet2.net,1)*sim(memfunet2.net,1)+sim(memfynet3.
net,1)*sim(memfunet3.net,1));
```

3) Program for the sliding mode output tracking controller

```
function [sys,x0,str,ts]=sfSMOTC(t,x,u,flag)
switch flag,
case 0     % initializations
  [sys,x0,str,ts] = mdlInitializeSizes;
case 2     % discrete states updates
  sys = mdlUpdates(x,u);
case 3     % computation of control signal
  sys=mdlOutputs(t,x,u);
case {1, 4, 9} % unused flag values
  sys = [];
otherwise    % error handling
  error(['Unhandled flag = ',num2str(flag)]);
end;
```

```
%==============================================================
% when flag=0, perform system initialization
%==============================================================
function [sys,x0,str,ts] = mdlInitializeSizes
sizes = simsizes;     % read default control variables
sizes.NumContStates = 0; % no continuous states
sizes.NumDiscStates = 3; % 3 states
sizes.NumOutputs = 1;  % 1 output variables: control u(t) and state x(3)
sizes.NumInputs = 6;   % 5 input signals
sizes.DirFeedthrough = 1;% input reflected directly in output
sizes.NumSampleTimes = 1;% single sampling period
sys = simsizes(sizes); %
x0 = [0; 0; 0];      % zero initial states
str = [];
ts = [-1 0];        % sampling period

%==============================================================
% when flag=2, updates the discrete states
%==============================================================
function sys = mdlUpdates(x,u)
T=0.1;

sys=[(u(1)-u(2))/T;
     (u(2)-u(3))/T;
     (u(4)-u(5))/T];

%==============================================================
% when flag=3, computates the output signals
%==============================================================
function sys = mdlOutputs(t,x,u)
T=0.1;

r=u(1);
r_1=u(2);
r_2=u(3);
dr=x(1);
dr_1=x(2);
e=u(4);
de=x(3);

c=30;
Ce=[c,1];
```

```
ul=u(6);

A2=0.8602;
B2=1.6319;

%Using Waitui method
r1=2*r-r_1;
dr1=2*dr-dr_1;

R=[r;dr];
R_1=[r_1;dr_1];
R1=[r1;dr1];
fk=R1(1)-A2*R(1);
fk_1=R(1)-A2*R_1(1);

s=c*e+de;
Ae=A2;
Be=-B2;
f0=2.2*2*abs(s)/(abs((c+1)*Be)*(abs(e)+abs(de)));
deta1=0.1*f0*((c+1)*Be)*((c+1)*Be)*abs(e)*(abs(e)+abs(de));
deta2=0.1*f0*((c+1)*Be)*((c+1)*Be)*abs(de)*(abs(e)+abs(de));

cond1=(c+1)*Be*s*e;
if cond1<-deta1
  f1=f0;
elseif abs(cond1)<=deta1
  f1=0;
elseif cond1>deta1
  f1=-f0;
end

cond2=(c+1)*Be*s*de;
if cond2<-deta2
  f2=f0;
elseif abs(cond2)<=deta2
  f2=0;
elseif cond2>deta2
  f2=-f0;
end
Fd=[f1,f2];
ueq=(c*(1-Ae)*e+(1-Ae)*de-c*fk+Be/T*ul-(fk-fk_1)/T)/(c*Be+Be/T);
ut=ueq+Fd*[e;de];

sys=ut;
```

References

[1] Yi, J. Q., N. Yubazaki and K. Hirota. 2001. Upswing and stabilization control of inverted pendulum system based on the SIRMs dynamically connected fuzzy inference model. Fuzzy Sets and Systems. 122: 139–152.

[2] Yi, J. Q., N. Yubazaki and K. Hirota. 2001. Stabilization control of series-type double inverted pendulum systems using the SIRMs dynamically connected fuzzy inference model. Artificial Intelligence in Engineering. 15: 297–308.

[3] Yi, J. Q., N. Yubazaki and K. Hirota. 2002. A proposal of SIRMs dynamically connected fuzzy inference model for plural input fuzzy control. Fuzzy Sets and Systems. 125: 79–92.

[4] Yubazaki, N., J. Q. Yi, M. Qtani, N. Unemura and K. Hirota. 1997. Trajectory traking control of unconstrained objects based on the SIRMs dynamically connected fuzzy inference model. In: Proceedings FUZZ-IEEE'97. Barcelona, Spain. 1–5 July 1997. IEEE. Piscataway, USA.

[5] Tong, S. W. and D. W. Qian. 2013. Control of a fuel cell based on the SIRMs fuzzy inference model. International Journal of Hydrogen Energy. 38: 4124–4131.

[6] Baba, N. 1989. A new approach for finding the global minimum of error function of neural networks. Neural Networks. 2: 367–373.

[7] Press, W. H., S. A. Teukolsky, W. T. Vetterling and B. P. Flannery. 1992. Numerial recipes in C. Cabridge University Press. Oxford, UK.

[8] Tong, S. W., J. J. Fang and Y. N. Zhang. 2017. Output tracking control of a hydrogen-air PEM fuel cell. IEEE/CAA Journal of Automatica Sinica. 4: 273–279.

[9] Tong, S. W., D. W. Qian and J. J. Fang. 2015. Sliding mode output tracking control based on a fuzzy clustering model. In: Proceedings of the 2015 International Conference on Advanced Mechatronic Systems, Beijing, China. 22–24 August 2015.

[10] Gustafson, D. E. and W. C. Kessel. 1979. Fuzzy clustering with a fuzzy covariance matrix. In: Decision and Control including the 17th Symposium on Adaptive Processes. San Diego, USA, 10–12 January 1979, IEEE. San Diego, USA.

6 Conclusion and future work

6.1 Conclusion

In this book, the control methods of fuel cell without and with load current feedback are studied for 1 kW hydrogen-air proton exchange membrane fuel cell in the Institute of Automation, Chinese Academy of Sciences. Model-based and data-based control approaches are all considered. Specific conclusions can be drawn as discussed in the following section.

In the study of no load current feedback fuel cell control method, starting from the linear control method, an adaptive state feedback predictive control method based on SPM model has been proposed. The SPM model was used to convert the nonlinear problem of parameter and state multiplication into the linear estimation problem. Combined with the extended Kalman filter, the parameters and states can be simultaneously estimated to design a state feedback predictive controller for the case where the noise information is unknown. The control method is better than the traditional control approach since more controlled process information is considered. Using C language single-valued predictive control algorithm, the calculation is simple and more suitable for real-time control. The SPM-based adaptive state feedback predictive control algorithm can better realize the pressure balance control between the anode and cathode of the proton exchange membrane fuel cell.

Since the fuel cell is essentially a complex nonlinear time-varying process, it is difficult to establish an accurate linear model. Using nonlinear model-free method will further improve the control performance. Therefore, in the study of no load current feedback fuel cell control method, this book proposed a real-time simplified variable domain fuzzy control algorithm. Combining the simplified fuzzy reasoning and variable domain theory, the control algorithm not only can be suitable for real-time control but also can achieve precise control. Through a simplified fuzzy reasoning method, it only needs to consider the relevant control rules in each reasoning process. For example, for the particular triangular membership functions of the two-input, one-output fuzzy controller, only four control rules are activated during each inference process, which makes the design of the controller very easier. By introducing the idea of variable domain, the domain contracts and expands with the changes of the error. The contraction of the domain is equivalent to the increase of the control rules, so that the limited division of the membership can be used to achieve more precise control. Real-time fuzzy reasoning and variable domain ideas are perfectly integrated into the proton exchange membrane fuel cell for cathode and anode pressure balance control, not only meeting real-time requirements of the hardware, but also getting better control accuracy than other linear methods.

When studying the control method of fuel cells with load current feedback, the single-input rule module's dynamic connected fuzzy inference model is applied for

https://doi.org/10.1515/9783110602159-006

the first time to realize the control of two controlled variables (cathode pressure and air flow) with an operating variable (the given voltage for the air mass flow meter). In order to meet the need of online adjustment of model parameters, an online random search optimization method based on C-S function is also designed and implemented in the control of the fuel cell device. To deal with the soft characteristic of the stand-alone fuel cell stack output, an output tracking control algorithm is designed. The method utilizes the fuzzy cluster modeling technique to transform the nonlinear fuel cell system into a T-S model that uses three individual local models for approximation by overlapping membership functions to realize smooth transition of these submodels. Neural networks are substituted for the implicit membership functions. Thus, the sliding mode control law can be calculated directly. The sliding mode control together with the integral action can realize good agreement between the reference stack voltage and the actual stack output.

Although the control method proposed in this book is for the actual 1 kW hydrogen proton exchange membrane fuel cell device of the Institute of Automation, Chinese Academy of Sciences, it is not limited to this situation. We can also apply it to other similar situations or controlled processes.

6.2 Future work

In this book, a series of new and practical control methods are proposed and applied for a 1 kW hydrogen-air proton exchange membrane fuel cell experimental device of the Institute of Automation, Chinese Academy of Sciences, and a lot of simulation and experiment work is being done. It should be noted that there is still a lot of work to be done, which is summarized as follows:

1) Stability is a very important problem in the design of control system. There is no stability analysis of the proposed control methods, such as real-time simplified variable domain fuzzy control and SPM-based adaptive state feedback predictive control. In the future, research work needs to further elaborate on this issue.
2) The work in this book is mainly focused on the control of the gas supply subsystem. Other subsystems, such as humidifying subsystem and power conditioning subsystem, also deserve deep study. Since the fuel cell is a system composed with several subsystems, the efficiency and optimization of the whole devices rely on the coordination between these subsystems.
3) In this book, only the stand-alone fuel cell is investigated in the scope of producing electricity by a hybrid system with fuel cell, supercapacitor, battery, or other auxiliary energy sources. The energy management and optimization are also a very meaningful and promising research direction.

In short, the fuel cell is a very complex nonlinear system that requires multidisciplinary cross-research, from whole to partial, and then from partial to integral. With the deepening of the study of fuel cells, there is reason to believe that the fuel cell will become the main application of the form of hydrogen technology in the twenty-first century.

Index

https://doi.org/10.1515/9783110602159-007

Bei Fragen zur Produktsicherheit wenden Sie sich bitte an:
If you have any questions regarding product safety,
please contact:

Walter de Gruyter GmbH
Genthiner Straße 13
10785 Berlin
productsafety@degruyterbrill.com